Explorers Of
AUSTRALIA

Captain Charles Sturt, aged about 54 years. From the painting by Crossland.

EXPLORERS OF AUSTRALIA

ERNEST FAVENC

SENATE

Explorers of Australia

First published in 1908 by Whitcombe & Tombs Ltd,
Christchurch, Melbourne, and London

This edition published in 1998 by Senate,
an imprint of Tiger Books International plc,
26A York Street, Twickenham,
Middlesex TW1 3LJ, United Kingdom

Cover design © 1998 Tiger Books International

1 3 5 7 9 10 8 6 4 2

All rights reserved. No part of this publication may be
reproduced, stored in a retrieval system or transmitted,
in any form or by any means, electronic, mechanical,
photocopying, recording or otherwise, without the prior
permission of the copyright owners.

ISBN 1 85958 537 X

Printed and bound in the UK by
Cox & Wyman, Reading, England

AUTHOR'S PREFACE.

In presenting to the public this history of those makers of Australasia whose work consisted in the exploration of the surface of the continent of Australia, I have much pleasure in drawing the reader's attention to the portraits which illustrate the text. It is, I venture to say, the most complete collection of portraits of the explorers that has yet been published in one volume. Some of them of course must needs be conventional; but many of them, such as the portrait of Oxley when a young man, and of A. C. Gregory, have never been given publicity before; and in many cases I have selected early portraits, whenever I had the opportunity, in preference to the oft published portrait of the same subject when advanced in years.

There are many who assisted me in the collection of these portraits. To Mr. F. Bladen, of the Public Library, Sydney; Mr. Malcolm Fraser, of Perth, W.A.; Mr. Thomas Gill, of Adelaide; Sir John Forrest; The Revd. J. Milne Curran; Mr. Archibald Meston; and many others my best thanks are due. In fact, in such a work as this, one cannot hope for success unless he seek the assistance of those who remembered the explorers in life, or have heard their friends and relatives talk familiarly of them. Let me particularly hope that from these pages our youth, who should be interested in the exploration of their native land, will form an adequate idea of the character of the men who helped to make Australia, and of some of the adverse conditions against which they struggled so nobly.

ERNEST FAVENC.

Sydney, 1908.

BIBLIOGRAPHY.

The published Journals of all the Explorers of Australia.
Reports of Explorations published in Parliamentary Papers.
History of New South Wales, from the Records. (Barton and Bladen.)
Account of New South Wales, by Captain Watkin Tench.
Manuscript Diaries of Blaxland, Lawson and Wentworth.
Manuscript Diaries of G. W. Evans. (Macquarie and Lachlan Rivers.)
The Pioneers of Victoria and South Australia, by various writers.
Contemporaneous Australian Journals of the several States.
Private letters and memoranda of persons in all the States.
Manuscript Diary of Charles Bonney.
Pamphlets and other bound extracts on the subject of exploration.
The Year Book of Western Australia.
Records of the Geographical Societies of South Australia and Victoria.
Russell's Genesis of Queensland.
Biographical Notes, by J. H. Maiden.
Spinifex and Sand, by David Carnegie.

INTRODUCTION.

In introducing this book, I should like to commend it to its readers as giving an account of the explorers of Australia in a simple and concise form not hitherto available.

It introduces them to us, tells the tale of their long-tried patience and stubborn endurance, how they lived and did their work, and gives a short but graphic outline of the work they accomplished in opening out and preparing Australia as another home for our race on this side of the world.

The battle that they fought and won was over great natural difficulties and obstacles, as fortunately there were no ferocious wild beasts in Australia, while the danger from the hostility of the aborigines (though a barbarous people) was with care and judgment, with a few exceptions, avoided.

Their triumph has resulted in peaceful progress and in permanent occupation and settlement of a vast continent.

Of all the Australian explorers the fate of Leichhardt—"the Franklin of Australia," as the author so justly terms him—is alone shrouded in mystery. "No man knoweth his sepulchre to this day." His party of six white men (including Leichhardt) and two black boys, with 12 horses, 13 mules, 50 bullocks, and 270 goats, have never been heard of since they left McPherson's station on the Cogoon on 3rd April, 1848; and although there have been several attempts to unravel the mystery, there is scarcely a possibility of any discovery in regard to their fate ever being made.

There can be no doubt that the fascination concerning the work of the early explorers of Australia will "gather strength as it goes." Hitherto we have been too close to them rightly to appreciate what was done. This book therefore comes at an opportune time, and is a valuable record. The author has already done a great service to Australian explorations by his writings, and in the present instance has added to our obligation to him by condensing the records into a smaller compass, and by that means has brought it within convenient limits for use in schools and for general readers.

Of the explorers of Australia, eleven have been honoured, by being placed on the "Golden Roll" (Gold Medallists) of the Royal

Geographical Society of London; Edward John Eyre being the first to receive the honour in 1843, and Ernest Giles being the eleventh and last to receive it in 1880. In the order of Nature "one generation passeth away and another generation cometh," and so it comes to pass that every one on the "Golden Roll" except myself has gone to "the undiscovered country from whose bourn no traveller returns."

That the Australian people will always remember the deeds of those, who, in their day and generation, under arduous and difficult conditions devoted themselves to the exploration of the Continent goes without saying, and I, who in bygone years had the honour of assisting in the task, heartily wish that such fruit may be born of those deeds that Australia will continue to increase and flourish more and more abundantly, and thus fulfil her destiny as the great civilising and dominating power in the Southern Seas.

JOHN FORREST.

"The Bungalow,"
    Hay Street, Perth,
        Western Australia,
January 7th, 1908.

CONTENTS.

	PAGE
PREFACE	vii.
BIBLIOGRAPHY	viii.
INTRODUCTION, by Sir John Forrest	ix.
CONTENTS	xi.
LIST OF ILLUSTRATIONS	xiii.

PART I.—EASTERN AUSTRALIA.

CHAPTER.	
I. ORIGINS	1
(i.) Governor Phillip	1
(ii.) Captain Tench	4
(iii.) The Blue Mountains: Barallier	4
(iv.) The Blue Mountains: Blaxland	7
II. GEORGE WILLIAM EVANS	16
(i.) First Inland Exploration	16
(ii.) The Lachlan River	19
(iii.) The Unknown West	21
III. JOHN OXLEY	23
(i.) General Biography	23
(ii.) His First Expedition	26
(iii.) The Liverpool Plains	32
(iv.) The Brisbane River	39
IV. HAMILTON HUME	42
(i.) Early Achievements	42
(ii.) Discovery of the Hume (Murray)	44
V. ALLAN CUNNINGHAM	50
(i.) Coastal Expeditions	50
(ii.) Pandora's Pass	52
(iii.) The Darling Downs	53
VI. CHARLES STURT	59
(i.) Early Life	59
(ii.) The Darling	60
(iii.) The Passage of the Murray	67
VII. SIR THOMAS MITCHELL	76
(i.) Introductory	76
(ii.) The Upper Darling	78
(iii.) The Passage of the Darling	80
(iv.) Australia Felix	84
(v.) Discovery of the Barcoo	88
VIII. THE EARLY FORTIES	93
(i.) Angas McMillan and Gippsland	93
(ii.) Count Strzelecki	94
(iii.) Patrick Leslie	95
(iv.) Ludwig Leichhardt	95

CONTENTS

CHAPTER		PAGE
IX.	EDMUND B. KENNEDY	110
	(i.) The Victoria River and Cooper's Creek	110
	(ii.) A Tragic Expedition	112
X.	LATER EXPLORATION IN THE NORTH-EAST	122
	(i.) Walker in Search of Burke and Wills	122
	(ii.) Burdekin and Cape York Expeditions	123

PART II.—CENTRAL AUSTRALIA.

XI.	EDWARD JOHN EYRE	135
	(i.) Settlement of Adelaide and the Overlanders	135
	(ii.) Eyre's Chief Journeys	139
XII.	ATTEMPTS TO REACH THE CENTRE	151
	(i.) Lake Torrens Pioneers and Horrocks	151
	(ii.) Charles Sturt	153
XIII.	BABBAGE AND STUART	169
	(i.) B. Herschel Babbage	169
	(ii.) John McDouall Stuart	175
XIV.	BURKE AND WILLS	186
XV.	BURKE AND WILLS RELIEF EXPEDITIONS AND ATTEMPTS TOWARDS PERTH	201
	(i.) John McKinlay	201
	(ii.) William Landsborough	206
	(iii.) Major P. E. Warburton	208
	(iv.) William Christie Gosse	213
XVI.	TRAVERSING THE CENTRE	216
	(i.) Ernest Giles	216
	(ii.) W. H. Tietkins and Others	222

PART III.—WESTERN AUSTRALIA.

XVII.	ROE, GREY, AND GREGORY	233
	(i.) Roe and the Pioneers	233
	(ii.) Sir George Grey	237
	(iii.) Augustus C. Gregory	242
XVIII.	A. C. AND F. T. GREGORY	247
	(i.) A. C. Gregory on Sturt's Creek and the Barcoo	247
	(ii.) Frank T. Gregory	253
XIX.	FROM WEST TO EAST	264
	(i.) Austin	264
	(ii.) Sir John Forrest	267
	(iii.) Alexander Forrest	279
XX.	LATER WESTERN EXPEDITIONS	281
	(i.) Cambridge Gulf and the Kimberley District	281
	(ii.) Lindsay and the Elder Exploring Expedition	282
	(iii.) Wells and Carnegie in the Northern Desert	285
	(iv.) Hann and Brockman in the North-West	292
INDEX OF NAMES OF PERSONS		297
INDEX OF PLACE NAMES		300

ILLUSTRATIONS.

	PAGE
Charles Sturt	*Frontispiece*
Gregory Blaxland	9
George William Evans	16
John Oxley	25
Lachlan River	31
Hamilton Hume	42
Allan Cunningham	50
The Cunningham Memorial, Sydney	57
Darling River	65
Junction of Darling and Murray Rivers	71
Sir Thomas Mitchell	76
A Chief of the Bogan River Tribe	81
Ludwig Leichhardt	96
John Frederick Mann	105
Edmund B. Kennedy	110
Wild Blacks of Cape York	117
Frank Jardine	124
Alec Jardine	126
John McDouall Stuart	134
Edward John Eyre	139
John Ainsworth Horrocks	152
Sturt's Depot Glen	159
Poole's Grave and Monument	163
B. Herschel Babbage	169
John McDouall Stuart	175
Robert O'Hara Burke	186
William John Wills	187
Scenes on Cooper's Creek (*Howitt*)	191
John King	195
Edwin J. Welch	196
Burke and Wills Monument, Melbourne	199
Major P. E. Warburton	209
William Christie Gosse	214
Baron Sir Ferdinand von Mueller	217
Caravan of Camels in an Australian Desert	219
W. H. Tietkins	222
Ernest Favenc	225
John Septimus Roe	232
Sir George Grey	237
Rock Painting	239
Augustus C. Gregory	243
Frank T. Gregory	254

		PAGE
Maitland Brown	...	261
Sir John Forrest (1874)	...	268
Members of Geraldton—Adelaide Exploring Expedition, 1874	...	275
Alexander Forrest	...	279
W. Carr-Boyd	...	281
Sir Thomas Elder	...	283
David Lindsay	...	284
L. A. Wells	...	286
David Wynford Carnegie	...	289
Frank Hann	...	292
Aboriginal Rock Painting, Glenelg River	...	294
Typical Australian Explorers of the Early Twentieth Century	...	295
Ernest Giles	...	296

MAPS AND PLANS.

1.—Routes of Blaxland, Wentworth, and Lawson (1813); Evans (1813); Oxley (1817, 1818, 1823); and Sturt (1828-9)	11
2.—Routes of Hume and Hovell (1824); Sturt (1829-30); and Mitchell (1836)	45
3.—Routes of Sturt (1829-30); and Hume and Hovell (1824)	61
4.—Routes of Leichhardt (1844-5); Mitchell (1845-6); and Kennedy (1847-8)	97
5.—Routes of Eyre (1840-1)	141
6.—Basin of Lake Torrens, supposed extent and formation of	151
7.—Route of Sturt's Central Australian Expedition (1844-6)	155
8.—Routes of Stuart (1858-62); and Burke and Wills (1860-1)	177
9.—Routes of Grey (1836-7-9); Forrest (1869, 1870, 1874, 1879); and Giles (1873)	271

Part I.

EASTERN AUSTRALIA.

ERRATA.

Page 26, line 5, for "Frazer" read "Fraser."

Page 46, line 18, for "a few bushmen" read "some bush utensils."

Chapter I.

ORIGINS.

(i.)—Governor Phillip.

Arthur Phillip, whose claim to be considered the first inland explorer of the south-eastern portion of Australia rests upon his discovery of the Hawkesbury River and a few short excursions to the northward of Port Jackson, had but scant leisure to spare from his official duties for extended geographical research. For all that, Phillip and a few of his officers were sufficiently imbued with the spirit of discovery to find opportunity to investigate a considerable area of country in the immediate neighbourhood of the settlement, and, considering the fact that all their explorations at the time had to be laboriously conducted on foot, they did their work well.

The first excursion undertaken by Phillip was on the 2nd of March, 1788, when he went to Broken Bay, whence, after a slight examination, he was forced to return by the inclemency of the weather. On the 15th of April he made another attempt to ascertain the character and features of the unknown land that he had taken possession of. Landing on the shore of the harbour, a short distance from the North Head, he started on a tour of examination, and, in the course of his march, penetrated to a distance of fifteen miles from the coast. At this point he caught sight of the distant range that was destined to baffle for many years the western progress of the early settlers. Phillip, on this his first glimpse of it, christened the northern elevations the Caermarthen Hills, and the southern elevations the Lansdowne; and a remarkable hill, destined to become a well-known early landmark, he called Richmond Hill. In the brief view he had of this range, there was suddenly born in Phillip's mind the conviction that a large river must have its source therein, and that upon the banks of such a river, the soil would be found more arable than about the present settlement. He at once made up his mind to try and gain the range on a different course.

A week later he landed at the head of the harbour and directed his march straight inland, hoping to reach either the mountains, which he knew to be there, or the river in whose existence he firmly believed. Disappointment dogged his steps; on the first day a belt of dense scrub forced his party to return and when, on the morrow, they avoided the scrub by following up a small creek and got into more thinly timbered country, their slow progress enabled them to accomplish only thirty miles in five days. By that time, they were short of provisions; there was no river visible, and the range still looked on them from afar. What cheered them was the sight of some land that promised richly to reward the labour of cultivation.

It was not until the 6th of June, 1789, that Phillip resumed his labours in the field of exploration. The "Sirius" had then returned from the Cape of Good Hope, and he could reckon on the assistance of his friend, Captain Hunter, to re-investigate Broken Bay with the vessel's boats. Accordingly, two boats were sent on to Broken Bay with provisions, where they were joined by the Governor and his party, who had marched overland. Besides Phillip, the party consisted of Captain Hunter and two of his officers, Captain Collins, Captain Johnston, and Surgeon White.

For two days they were engaged in examining the many inlets and openings of the Bay, and on the third, they chanced upon a branch that had before escaped their notice. They proceeded to explore it, and found the river of which Phillip had dreamed. The next day, renewed examination proved that it was indeed a noble river, with steep banks and a depth of water that promised well for navigation.

After their return to Sydney Cove, preparations were at once made to follow up this important discovery. On the 28th of June, Phillip, again accompanied by Hunter, left the Cove, having made much the same arrangements as before. There was a slight misunderstanding with regard to meeting the boat; but, after this was cleared away, the party soon floated out on to the waters of the new-found river. They rowed up the river until they reached the hill that Phillip, at a distance, had christened Richmond Hill. On traversing a reach of the stream, the main range, that as yet they had only dimly seen in the distance, suddenly loomed ahead of them, frowning in rugged grandeur close upon them, as it seemed. Struck with admiration and astonishment at this unexpected revelation of the deep ravines and stern and gloomy gorges that scored its front, over which hung a blue haze, Phillip, almost involuntarily, named them on the moment, the Blue Mountains. Next morning the explorers ascended Richmond Hill, from whose crest they looked across a deep, wooded valley to the mountains still many miles away. After a hasty examination of the country on the banks of the river, Phillip and his band returned to the settlement, he having now realised his brightest hopes and anticipations.

On the 11th of April, 1791, Phillip again started on an expedition, the object of which was a closer inspection of the Blue Mountains. He was accompanied this time by Captain Tench and Lieutenant Dawes; the latter, in December, 1789, had been sent out with a small party to reach the foot of the range, but had succeeded in approaching only within eleven miles of the Mountains, whence he was forced to retire by the rugged and broken nature of the country. On the present occasion, they reached the river two days after leaving Rose Hill. They followed it for another two days, but made no further discoveries, being greatly delayed by the constant detours around the heads of small tributary creeks, too deep to cross in the neighbourhood of the river.

This was the last exploring expedition undertaken by Governor Phillip. Considering that his health was not robust, and that the work entailed was of a specially arduous nature, his personal share in exploring the country about the little settlement was noteworthy. It proved him to possess both the foresight and the energy necessary in an explorer.

(ii.)—Captain Tench.

In the month of June, 1789, Captain Watkin Tench, who, during his short sojourn in the infant colony showed himself as zealous in exploration as he was keen in his observations, started from the newly-formed redoubt at Rose Hill, of which he was in command, on a short excursion to examine the surrounding country. This trip, inspired by Tench's ardent love of discovery, became a noteworthy one in the annals of New South Wales. It was made during the month that witnessed the discovery of the Hawkesbury River. On the second day after his party left Rose Hill, they found themselves early in the morning on "the banks of a river, nearly as broad as the Thames at Putney, and apparently of great depth, the current running very slowly in a northerly direction."

This river, at first known as the Tench, was afterwards named the Nepean by Phillip, when its identity as a tributary of the Hawkesbury had been confirmed. Two other slight excursions were made by Tench in company with Lieutenant Dawes, who was in charge of the Observatory, and ex-surgeon Worgan. In May, 1791, Tench and Dawes started from Rose Hill and confirmed the supposition that the Nepean was an affluent of the Hawkesbury, a matter over which there had been some doubt since its first discovery by Tench. Tench returned to England in H.M.S. "Gorgon," in December, 1791.

The names of Paterson, Johnson, Palmer, and Laing are also connected with exploration on the upper Hawkesbury.

(iii.)—The Blue Mountains: Barallier.

The exploration of that portion of Australia which was accessible by the scanty means of the early settlers was for many years impeded by the stern barrier of the mountains, and most of their efforts in the direction of discovery were aimed at surmounting the range that defied their attacks. Among the many whose attempts were signalised only by failure were the gallant Bass, whose name, for other reasons, will never be forgotten by Australians, the quarrelsome and pragmatic Cayley, and the adventurous Hack. Amongst them there was one, however, whose failure, read by the light of modern knowledge, was probably a geographical success. This was Francis Barallier, ensign in the New South Wales corps, who was encouraged by Governor King to indulge his ardent longing for discovery. By birth a Frenchman, Barallier had received his ensigncy by commission on the 13th of February, 1801, having done duty as an ensign since July, 1800, by virtue of a government general order issued by Governor Hunter. In August, 1801, he had been appointed by Governor King military engineer, in place of Captain Abbott resigned. In February, 1802, he was succeeded by Lieutenant George Bellasis, an artillery officer. Besides his expeditions to the Blue Mountains, he did much surveying with Lieutenant James Grant in the "Lady Nelson." In 1804, he went to England and saw service in several regiments, distinguishing himself greatly in military engineering, amongst his works being the erection of the Nelson Column in Trafalgar Square, the designer of which was Mr. Railton. Barallier died in 1853.

Peron, the French naturalist, tells us that when in Sydney in October, 1802, he persuaded Governor King to fit out a party to attempt the passage of the mountains, and that a young Frenchman, aide-de-camp to the Governor, was intrusted with the leadership. He returned, however, without having been able to penetrate further than his English predecessors.

On the following month, however, Barallier set out from Parramatta, on his famous embassy to the King of the Mountains. This fictitious embassy arose from the fact that Colonel Paterson having refused Barallier the required leave, King claimed him as his aide-de-camp, and sent him on this embassy. Barallier started with four soldiers, five convicts, and a waggon-load of provisions drawn by two bullocks. He crossed the Nepean and established a depot at a place known as Nattai, whence the waggon was sent back to Sydney for provisions, Barallier, with the remainder of his men and a native, pushing out westwards. After this preliminary examination he returned to the depot, and made a fresh departure on the 22nd of November, and, continuing mostly directly westwards, he reached a point (according to his chart) about one hundred and five miles due west from Lake Illawarra. If this position is even approximately correct, he must have been at the very source of the Lachlan River.

I give a few extracts from his diary, which was not even translated until the Historical Records of New South Wales were collected by Mr. F. M. Bladen. They refer to the crossing of the range.

"On the 24th of November, I followed the range of elevated mountains, where I saw several kangaroos. This country is covered with meadows and small hills, where trees grow a great distance apart. . . I resumed my journey, following various directions to avoid obstacles, and at 4 o'clock I arrived on the top of a hill where I discovered that the direction of the chain of mountains extended itself north-westerly to a distance which I estimated to be about thirty miles, and which turned abruptly at right angles. It formed a barrier nearly north and south, which it was necessary to climb over. . . At 7 o'clock I arrived on the summit of another hill, from where I noticed three openings: the first on the right towards N. 50. W.; the other in front of me, and which appeared very large, was west from me; and the third was S. 35. W.

. . . This discovery gave me great hope, and the whole of the party appeared quite pleased, thinking that we had surmounted all difficulties, and that we were going to enter a plain, the apparent immensity of which gave every promise of our being able to penetrate far into the interior of the country. . . At six o'clock I found

myself at a distance of about two miles from the western
passage. . . I was then only half-a-mile from the
passage, and I sent on two men in order to discover it,
instructing them to ascend the mountain to the north of
this passage. . . I waited till 7 o'clock for my two
men, who related to me, that after passing the range
which was in front of us we would enter an immense
plain, that from the height where they were on the mountain, they had caught sight of only a few hills standing
here and there on this plain, and that the country in front
of them had the appearance of a meadow. . . At
daybreak I left with two men to verify myself the configuration of the ground, and to ascertain whether the
passage of the Blue Mountains had really been effected.
I climbed the chain of mountains north of us. When I
had reached the middle of this height the view of a plain
as vast as the eye could reach confirmed to me the report
of the previous day. . . I discovered towards the west
and at a distance which I estimated to be forty miles,
a range of mountains higher than those we had passed.
. . . From where I was, I could not detect any obstacle
to the passage right to the foot of those mountains. . .
After having cut a cross of St. Andrew on a tree to
indicate the terminus of my second journey, I returned
by the same route I had come.''

Barallier concludes his diary by mentioning another
projected expedition over the mountains from Jervis
Bay. But no record of such a journey has ever come
to light.

(iv.)—The Blue Mountains: Blaxland.

Whether Barallier succeeded or not in reaching the
summit of the mountains, the verdict accepted at that
date was that they had not been passed; and until the
year 1813, they were regarded as impenetrable. The
narrative of the crossing of these mountains, and the
chain of events that led up to the successful attempt is
widely known, but only in a general way. It is for this
reason that a longer and more detailed account is given
in these pages; and as the expedition was successful in

opening up a way to the interior of the Continent, it is fitting that its leader and originator, Gregory Blaxland, should be classed amongst the makers of Australasia.

Blaxland was born in Kent, in 1771, and arrived in the colony in 1806, accompanied by his wife and three children. He settled down to the congenial occupation of stockbreeding, on what was then considered to be a large scale. Finding that his stock did not thrive so well in the immediate neighbourhood of the sea coast, and wanting more land for pasturing his increasing herds, he made anxious enquiries in all directions as to the possibility of crossing the Blue Mountains inland. Nobody would entertain such a suggestion, the failures had been too many: every one to whom he broached the subject declared it to be impossible, prophesying that the extension of the settlement westward would forever be obstructed by their unscalable heights. Blaxland, however, was not intimidated by these disheartening predictions; and, in 1811, he started out on a short journey of investigation, in company with three Europeans and two natives. On this trip he found that by keeping on the crowning ridge or dividing watershed between the streams running into the Nepean and those that fed what he then took to be an inland river, he got along fairly well. Some time afterwards he accompanied the Governor in a boat excursion up the Warragamba, a tributary of the Nepean, and though there were no noteworthy results, it convinced Blaxland that, could he follow his former tactics of adhering to the leading ridge that formed the divide between the tributaries of the northern bank of this river and the affluents of the Grose, a tributary of the Hawkesbury, he would attain his object and reach the highlands. It will thus be seen that Blaxland acted with a definite and well-thought-out mode of procedure; and that the ridge he selected for the attempt was chosen with judgment based on considerable knowledge of the locality, which he gained from many talks with the men who hunted and frequented the foothills of the range. Finally, when he had arranged his plan of assault,

Statue of Gregory Blaxland, Lands Office, Sydney.

he confided his intention to two friends, Lieutenant William Lawson and William Charles Wentworth, whose names are associated with his in the conquest of the Mountains. They both consented to accompany him, and agreed to follow his idea of stubbornly following one leading spur. Blaxland's former expedition had convinced him that the local knowledge of the natives did not extend far enough to be of any service, and they therefore did not take any aborigines with them. They took pack-horses, however, which proves that the party started with a well-founded faith in their ultimate success, and gave no heed to the terrifying descriptions of former travellers.

The besetting hindrance to their progress was the low scrub of brushwood that greatly delayed the pack-horses. This obstacle was overcome only by patiently advancing before the horses every afternoon, and cutting a bridle-track for the succeeding day's stage. Thus literally, the way that ultimately led into the interior was won foot by foot, and the little pioneering band eventually descended into open grazing country at the head of what is now known as the Cox River. The outward and return trip occupied less than one month's time; which speaks volumes for the wise choice of route; but what says more, is the fact that no better natural, upward pathway has since been found.

A synopsis of Blaxland's journal is given here, commencing with a few quoted lines of preamble:—

"On Tuesday, May 11th, 1813, Mr. Gregory Blaxland, Mr. William Wentworth and Lieutenant Lawson, attended by four servants, with five dogs and four horses laden with provisions and other necessaries, left Mr. Blaxland's farm at South Creek for the purpose of endeavouring to affect a passage over the Blue Mountains, between the Western River* and the River Grose. . . . The distance travelled on this and subsequent days was computed by time, the rate being estimated at about two miles per hour."

*The Warragamba.

ORIGINS 11

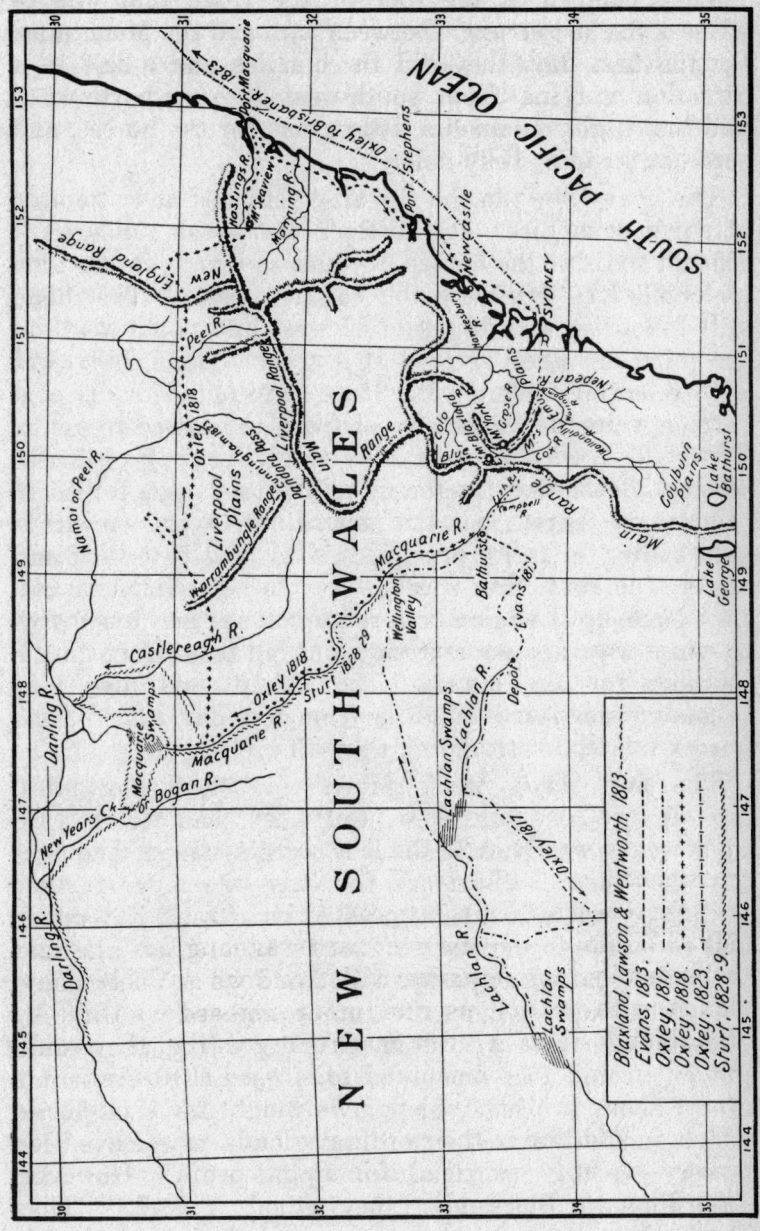

They camped at the foot of the ridge that was to witness the last struggle between man and the Mountains. On the first day, they did three miles and a-half in a direction varying from south-west to west-north-west, and that night obtained a little grass for the horses, and some water in a rocky hole.

The heavy dews in the morning retarded any attempts at early departures, as the thick wet brush rendered it difficult to drive the horses, so that, as a rule, it was nine o'clock before they were able to strike camp. The ridge, still favouring the direction of west and north-west, on the third day they arrived at a tract of land, hilly, but with tolerable grass on it. Here they found traces of a former white visitant in the shape of a marked-tree line. Two miles from this point, they met with a belt of brush-wood so dense that for the first time they were forced to alter their course; but the subordinate spurs on either side ending in rocky precipices, they had to return and again confront the scrub. In these circumstances, they made up their minds to rely upon axe and tomahawk to win a way, and so next morning fell to work cutting a passage for the horses. The ascent was also now becoming steep and rough, and on this day some of the horses fell while struggling up with their loads.

The first day's work gained for them five miles, but at the end of their toil they had to retrace their weary way back to the last night's camp. The next day they cleared the track for only two miles further ahead; so much time being wasted in walking backwards and forwards to the work. There was no grass amongst the scrub that encompassed them, and when, on Monday, they determined to move the camp equipage forward, they packed the horses with as much cut-grass as they could put on them. This amounted to, according to Lawson's diary, about two hundred pounds weight for each horse, which, in addition to their ordinary loads, must have been a very weighty packload for uphill work. However, according to Blaxland, "they stood it well." They obtained no water for their animals that night, and what they wanted for their own requirements had to be painfully carried up a cliff about six hundred feet in height. On the succeeding day they suddenly came on what at first appeared to be an impassable barrier. The ridge which they had so pertinaciously followed, had, for the last mile, narrowed and dwindled down into a sharp razor-backed spur, flanked with rugged and abrupt gullies on either slope. Across this narrow way now stretched a perpendicularly-sided mass of rock, which seemed effectually to bar their path. The removal of a few large boulders however, revealed an aperture which, after some labour, they widened sufficiently to allow the pack-horses to squeeze through.

Once through they began to ascend what they estimated to be the second tier of the Mountains. Shortly after they left camp that morning they came on a pile of stones, or cairn, evidently the work of some European, which they attributed to Bass. They were much elated at the thought that they had now passed beyond the limit of any previous attempt.*

They could now look round with triumph on the panorama spread beneath their view, and from the superior elevation which they had obtained, they took the bearings of several noticeable landmarks that they had seen before only from the flat country. The labour of cutting a path each day for the horses for the next day's march had, however, still to be continued; but the crest of the ridge was again wider, though the gullies on each side were as steep as before. That night, in camp, the dogs were uneasy throughout the night, and several times gave tongue and aroused the sleepers, tired with their day's work. From what they found afterwards, they had good reason to believe that the blacks had been lurking around meditating an attack.

They then passed over the locality known in the present day as Blackheath, and soon afterwards had their course diverted to the northward by what Blaxland terms

*This cairn was afterwards named " Cayley's Repulse " by Governor Macquarie: but recent research goes to show that Cayley followed the valley of the Grose, and was many miles to the north of where the cairn was found. According to Flinders, Bass was not on the high ridge traversed by Blaxland and party.

"a stone wall rising perpendicularly out of the side of the mountain." This they tried to descend, but without success, and so kept on along its brow. Undergrowth still delayed them, and they still had to spend their energies in hewing a passage, until on the 28th of the month, they camped on the edge of the steep descent that had lately marched beside them. The decline was, however, not quite so abrupt, and the face no longer composed of solid rock. They paused to overlook what lay before them and immediately below, and found the view more gratifying than they had anticipated. What they had at first taken for sandy barren soil proved now, on nearer inspection, to be forest-land fairly covered with a good growth of grass. The horses not having tasted fresh grass for some days, they cut a slanting trench across the sloping face of the descent in order to afford the horses some sort of foothold, and managed to get them down to a little feed that evening.

Next morning they were up and away early, and reached the foot of the mountain (Mount York) at 9 a.m., having had to carry the pack-loads down most of the way themselves, as it was too steep for laden horses to preserve their balance with safety. The actual base of the mountain was reached through a gap in the rocks, some thirty feet in width.

They now found themselves on what was then termed "meadow land," drained by the upper tributaries of the Warragamba; and this country presenting no serious obstacle to their further progress, they rightly concluded that they had now surmounted every difficulty. They followed the mountain stream up for some distance and, at the furthest point they reached, ascended a high sugarloaf hill, which surveyor Evans, who followed in their footsteps, called Mt. Blaxland. From the summit they had an extensive view all around, and Blaxland described the character of the country they saw in the following words:—"Forest and grass land, sufficient to support the stock of the colony for the next thirty years."

Just here, let us compare this prophecy with a similar one made by Evans a few months afterwards, on the pasture lands of the upper Macquarie:—"The increase of stock for some hundred years cannot overrun it."

The provisions of the explorers were now nearly expended; their apparel, especially their footgear, was in rags and tatters; on the other hand, the work that they had set themselves to do was well done. They had vanquished the Blue Mountains. Their return was uneventful. After breakfast on the 6th of June, they crossed the Nepean, their provisions, with the exception of a little flour, being quite consumed. We thus see how in the end the impenetrable range, that had so long overawed the colonists with its frown, was overcome, with slight difficulty, when local experience combined with method, was arrayed against it. To liken the former expeditions to Blaxland's is to compare a few headlong assaults with a well-conceived and skilfully worked-out attack. The men themselves write slightingly of the feat. Blaxland says:—"the passage of the Blue Mountains might be easily effected." Lawson's opinion of the mountain is:—"that there would be no difficulty in making a good road"; and Wentworth's verdict is:—"that the country they reached is easy of access." Evans, who was hot upon their trail, gives as his opinion:—"that there are no hills on the ridge that their ascent or descent is in any way difficult."

The tidings brought back by the party of successful pioneers created the greatest excitement in the little colony. No longer would the mountainous barrier stand defiantly in their western path. For over thirty years it had laughed at their puny efforts to cross its rugged crest, but its time had come at last; the way to the unknown west was now open, and rejoicingly the settlers prepared to follow on the explorers' trail. What the mysterious interior might hold, they could not imagine; but the gates thereto being thrown wide at last, its secrets would be soon known to them.

Blaxland died on the 3rd of January, 1853, having lived long enough to witness the wonderful advance in settlement due to his energies.

Chapter II.

GEORGE WILLIAM EVANS.

(i.)—First Inland Exploration.

George William Evans, Deputy-Surveyor of Lands, came forward at this stage as the most prominent figure in Australian exploration. To him is due the honour, without dispute or cavil, of being the first discoverer of an Australian river flowing into the interior.

George W. Evans, Discoverer of the Macquarie and Lachlan Rivers.

For some reason he has never received adequate recognition of his important explorations, and he is well-nigh forgotten by the people of New South Wales, the state that has benefited most by his labours. After Oxley's second expedition, his name appears to have been overshadowed by his official superior's. Yet his work was invariably successful, and his labour in the field unremitting.

Evans was born in England, at Warwick, in 1778. When a young man he went to the Cape of Good Hope, where he obtained an appointment in the dockyard, and while there he married his first wife, Janet Melvill. In 1802 he was appointed Deputy Surveyor-General, and came to Australia in H.M.S. "Buffalo," in order to take up his official duties. It was while he held this post that he carried out his work of exploration.

When he returned from these explorations, he resumed his duties as Deputy Surveyor-General only, until he was permanently settled in Tasmania, where he remained in office until the year 1825, when he resigned in disgust at his treatment by his superiors.

Evans lived at a time when official jealousies were rife, and men in position often heedless of the justice or veracity of their statements when influenced by party rancour. The machinations of a cabal led by Governor Arthur, and an effort made to deprive him of his well-deserved pension, necessitated Evans's departure for England to defend his claims. In this he was only partially successful, for the pension which it was understood was for life, was stopped in 1832. He returned to Tasmania, and passed the rest of his days at his residence, Warwick Lodge, at the head of Newtown Bay. He died at the age of seventy-four, and is buried in the old cemetery, Hobart; his second wife, Lucy Parris, rests in the same grave.

Evans was a clever draughtsman, and some of his sketches of the country explored are reproduced in Oxley's journal. He also published a book entitled "History and Description of the Present State of Van Diemen's Land."

It was on Saturday, the 20th of November, 1813, that Evans, in charge of five men, one of whom had been with Blaxland's party, started from the point of forest land on the Nepean known as Emu Island. He lost no time in following the tracks of the late expedition, leaving the measurement until his return. On Friday, the 26th, he reached Blaxland's furthest point, and thenceforward passed over new ground. It is somewhat amusing to note that his opinions of the country when on his outward way and on his homeward, are widely divergent. He candidly and ingenuously writes, after he has been on the table-land:—

"What appeared to me fine country on my first coming to it, looks miserable now after returning from so superior and good a country."

On Tuesday, the 30th of November, he gained a ridge that he had had in view for some time, though he had been "bothered" by the hills in his efforts to reach it. From this ridge he caught a tantalising view, a glimpse of the outskirts of the vast interior.

There before him, the first white man to look upon the scene, lay the open way to two thousand miles of fair pasture-lands and brooding desert-wastes—of limitless plains and boundless rolling downs—of open grassy forests and barren scrubs—of solitary mountain peaks and sluggish rivers; and, though then hidden from even the most brilliant imagination, the wondrous potentialities latent in that silent and untrodden region. If a vision of the future had been vouchsafed Deputy-Surveyor Evans as he stood and gazed—a vision of all that would cover the spacious lands before and beyond him before one hundred years had passed away—the entry he made in his diary would surely have reflected in its style his flight of imagination. Instead, we have the prosaic statement:—

"I came to a very high mount, when I was much pleased with the sight westward. I think I can see 40 miles which had the look of open country."

In a pleasant valley, he came upon a large "riverlett," and on its banks they camped. There they shot ducks and caught "trout"—as he called the Murray Cod—the first of the species to tickle the palate of a white man; fine specimens, too, weighing five and six pounds. As he proceeded further and further, he became enchanted with the scenery:—"The handsomest I have yet seen, with gently-rising hills and dales well-watered"—and he finally notes that language failed him to describe it adequately.

Evans named the river that led him through this veritable land of promise the Fish River, and a river which joined its waters with it from the south he called the Campbell River. The united stream he christened, as in duty bound, the Macquarie. Unimpeded in his course, he followed the Macquarie until he was 98½ measured miles—for they had been chaining since passing the limit of the first explorers—from the termination of Blaxland's journey. He then decided to return; for he had gained all the information he had been sent to seek;

and though game was plentiful, his party were without shoes, and the horses were suffering from sore backs.

Thus was concluded in a most satisfactory manner the first journey of exploration into the interior. Evans constantly saw, during his progress, unmistakeable traces of the natives; but he interviewed only a small party of five. This representative band of the inland aborigines of Australia, was composed of two lubras and some picaninnies, both the women being blind of the right eye.

The party reached the Nepean on their return journey on the 8th of January, 1814. Mr. Cox was immediately intrusted with the superintendence of the work of making a public road over the range, following closely the same route as that taken by Blaxland's party. This work was completed in the year 1815, and on the 26th of April of the same year, Governor Macquarie and a large staff set out to visit the newly-found territory. The Governor arrived at the recently-formed town of Bathurst on the 4th of May; but before his arrival Evans had been again ordered out on another exploring expedition to the south-west.

(ii.)—The Lachlan River.

Evans started from Bathurst on the 13th of May, 1815. He commenced his journey along the fine flat country then known as Queen Charlotte Vale, maintaining a southerly course for a day or two; but finding himself still amongst the tributaries of the Campbell River, he retraced his steps some twelve or fourteen miles in order to avoid a row of rocky hills. He then struck out more to the westward. On Thursday, the 23rd, he came to a chain of ponds bearing nearly north-west, and from a commanding ridge saw before him a prospect as gratifying as some of the scenes viewed on his former trip.

"I never saw a more pleasing-looking country. I cannot express the pleasure I feel in going forward. The hills we have passed are excellent land, well-wooded. To the south, distant objects are obscured by high hills, but in the south-west are very distant mountains, under them appears a mist as tho' rising from a river. It was the like look round to the west, but beyond the loom of high hills are very faintly distinguished."

This was the first view Evans obtained of the Lachlan valley. The ponds he had met with gradually grew into a connected stream: other ponds united with them from the north-east, and he writes:—"they have at the end of the day almost the appearance of a river." On the 24th he came to a creek which joined "the bed of a river rising in a N. 30. E. direction, now dry except in hollow places. It is fully 70 feet wide, having a pebbly bottom; on each side grow large swamp-oaks."

On Thursday, the 1st of June, this river holding a definite course to the westward, and he being clear of the points of the hills, which hitherto had hindered him greatly, he determined to return, as he was running short of provisions.

"To-morrow I am necessitated to return, and shall ascend a very high hill I left on my right hand this morning. I leave no mark here more than cutting trees. On one situated in an angle of the river on a wet creek bearing north I have deeply carved 'EVANS, 1st JUNE, 1815.' "*

On the next morning Evans ascended the hill he alluded to, and from the summit enjoyed a most extended view of the surrounding country, which he compared to a view of the ocean. On his way back to Bathurst, he bestowed upon the new river the name it now bears. A short passage in his diary, written during his return, is of peculiar interest, as it contains the first mention of snow seen in Australia by white men. On Thursday, the 8th of June, he writes:—

"The mountains I observed bearing north-west are covered with snow; I thought on my way out that their tops looked rather white. To-day it was distinguished as

---

*This tree, a tall and sturdy gum, flourished for over ninety years, and when in its prime was, unfortunately, owing to the spread of agricultural settlement, inadvertently ring-barked and killed. It must have been a fine tree when marked by the explorer, and though dead it is still standing at the date of the publication of this book. In 1906, the shield of wood bearing the inscription, was cut off by Mr. James Marsh, of Marshdale, and is now preserved in the Australian Museum in Sydney, N.S.W. It is the oldest marked-tree in the whole of Australasia.

plain as ever I saw snow on the mountains in Van
Diemen's Land. I never felt colder weather than it has
been some days past. We have broken ice full two inches
thick."

On the 12th of June the party returned to Bathurst,
and Evans had by that time accomplished two of the
most momentous journeys ever made in Australia. It was
not his actual discoveries alone that brought him fame,
but the vast field for settlement these discoveries opened
up. The independent explorations of Surveyor Evans
ceased after his discovery of the Lachlan; thenceforward
he served Australia as second to Lieutenant Oxley.

(iii.)—The Unknown West.

The settlers of that day took every advantage of the
new outlets for their energies, thrown open to them by the
recent successful explorations. Cattle and sheep were
rapidly driven forward on to the highlands, and,
favoured by a beautiful site, the town of Bathurst soon
assumed an orderly appearance. Private enterprise had
also been at work elsewhere. The pioneer settlers were
making their way south; the tide of settlement flowed over
the intermediate lands to the Shoalhaven River; and in
the north they had commenced the irresistible march of
civilization up the Hunter River.

It was in the Shoalhaven district that young Hamilton
Hume, the first Australian-born explorer to make his
mark in the field, gained his bushcraft.

Governor Macquarie, during his term of office,
did his best to foster exploration; and it was fortunate
that the first advance into the interior occurred when
there was a Governor in Australia who did not look coldly
upon geographical enterprise.

The men who entered first upon the task of solving the
geographical problems of the interior of the Australian
continent were doomed to meet with much bitter disappointment. The varying nature of the seasons caused the
different travellers to form contrary and perplexing
ideas, often with regard to the same tract of country.
What appeared to one man a land of pleasant gurgling

brooks, flowing through rich pastures, appeared to another as a pitiless desert, unfit for human foot to venture upon. Oxley, who traversed what is now the cream of the agricultural portion of the state of New South Wales, speaks of the main part of it in terms of the bitterest condemnation. His error was of course rather a mistake in judgment than the result of inaccurate observation.

Some of the colonists nursed far fonder hopes, and the general opinion seemed to be that these western-flowing rivers would gather in tributaries, and having swollen to a size worthy of so great a continent, seek the sea on the west coast. W. C. Wentworth, who certainly was capable of forming an opinion deserving consideration, wrote thus of the then untraced Macquarie River:—

"If the sanguine hopes to which the discovery of this river (the Macquarie) has given birth should be realised, and it should be found to empty itself into the ocean in the north-west coast, which is the only part of this vast island that has not been accurately surveyed, in what mighty conceptions of the future power and greatness of this colony may we not reasonably indulge? The nearest point at which Mr. Oxley left off to any part of the western coast is very little short of two thousand miles. If this river therefore be already of the size of the Hawkesbury at Windsor, which is not less than two hundred and fifty yards in breadth, and of sufficient depth to float a seventy-four gun ship, it is not difficult to imagine what must be its magnitude at its confluence with the ocean, before it can arrive at which it has to traverse a country nearly two thousand miles in extent. If it possesses the usual sinuosities of rivers, its course to the sea cannot be less than from five to six thousand miles, and the endless accession of tributary streams which it must receive in its passage through so great an extent of country will, without doubt, enable it to vie in point of magnitude with any river in the world."

It was to realise such ambitious hopes as these that Oxley went forth to penetrate into the interior.

Chapter III.

JOHN OXLEY.

(i.)—General Biography.

Oxley was born in England in the early part of 1781. In his youth he entered the navy, saw active service in many parts of the world, and rose to the rank of Lieutenant. He came to Australia in January, 1812, and was appointed Surveyor-General.

Throughout his career in Australia, Oxley would seem to have won the friendship and respect of all he came in contact with. Captain Charles Sturt, in the journal of his first expedition, wrote of him as follows:—

"A reflection arose to my mind, on examining these decaying vestiges of a former expedition, whether I should be more fortunate than the leader of it, and how far I should be able to penetrate beyond the point which had conquered his perseverance. Only a week before I left Sydney I had followed Mr. Oxley to the tomb. A man of great quickness and of uncommon ability. The task of following up his discoveries was no less enviable than arduous."

These thoughts were suggested to Sturt when standing at one of Oxley's old camps, and coming from such a man carry great weight.

The following obituary notice of Oxley appeared in the "Government Gazette" of May 27th, 1828.

"It would be impossible for his Excellency, consistently with his feelings, to announce the decease of the late Surveyor-General without endeavouring to express the sense he entertains of Mr. Oxley's services, though he cannot do justice to them.

"From the nature of this colony, the office of Surveyor-General is amongst the most important under Government; and to perform its duties in a manner Mr. Oxley has done for a long series of years is as honourable to his zeal and abilities as it is painful for the Government to be deprived of them.

"Mr. Oxley entered the public service at an early period of his life, and has filled the important situation of Surveyor-General for the last sixteen years.

"His exertions in the public service have been unwearied, as has been proved by his several expeditions to explore the interior. The public have reaped the benefit, while it is to be apprehended that the event, which they cannot fail to lament, has been accelerated by the privations and fatigues of these arduous services. Mr. Oxley eminently assisted in unfolding the advantages of this highly-favoured colony from an early stage of its existence, and his name will ever be associated with the dawn of its advancement. It is always gratifying to the Government to record its approbation of the services of meritorious public officers, and in assigning to Mr. Oxley's name a distinguished place in that class to which his devotion to the interests of the colony has so justly entitled him, the Government would do honour to his memory in the same degree as it feels the loss it has sustained in his death."

Oxley died at Kirkham, his private residence near Sydney, on the 25th of May, 1828. Though his judgment was at times at fault, as will be seen later on, he was essentially a successful explorer; for, although he did not in every case achieve the object aimed at, he always brought back his men without loss, and he opened up vast tracts of new country. John Oxley's personality is not very familiar, but the portrait presented to the reader in this volume was taken in the prime of his life, before he suffered the scars of doubtful battle with the Australian wilderness. It has never been published before, and is taken from the original miniature that he presented to Mrs. King, widow of Governor King, in 1810.

John Oxley. From a portrait in the possession of Mrs. Oxley, of Bowral. The portrait was presented to Mrs. King, widow of Governor King, in 1810, and signed by him.

(ii.)—His First Expedition.

On this, Oxley's first journey of exploration, Evans accompanied him as second in command, and another man who has left an immortal name was also with him— Allan Cunningham, officially known as King's Botanist. Charles Frazer, well-known in connection with the early history both of New South Wales and of Western Australia, accompanied Oxley under the title of Colonial Botanist. There were nine other men in the party—boatmen, horse-tenders, and so forth; they had with them two boats and fourteen pack and riding-horses. A depot was first formed at the junction of the small creek whence Evans had turned back, and where he had marked a tree with his initials in 1815. There the boats were launched and preparations completed for the final start. On the 6th of April, 1817, Oxley left Sydney and joined his party at the depot on the 1st of May. Thence he soon commenced this most momentous journey in Australia's early annals, eager to penetrate into the unknown, and inspired with hopes of solving the mystery of the outlet of this inland river.

Disappointment marks the tone of Oxley's journal from the start; the exceeding flatness of the country, the many ana-branches of the river, the low altitude of its banks, and the absence of any large tributary streams, above all, the dismal impression made by the monotony of the surroundings, seem to have depressed Oxley's spirit. He appears to have formed the idea that the interior tract he was approaching was nothing more than a dead and stagnant marsh—a huge dreary swamp, within whose bounds the inland rivers lost their individuality and merged into a lifeless morass. A more melancholy picture could not be imagined, and with such an awesome thought constantly haunting his mind there is no wonder that he became morbid, and that the dominant tone of his journal, whilst on the Lachlan, is so hopelessly pessimistic.

"These flats," he says, "are certainly not adapted for cattle; the grass is too swampy, and the bushes, swamps, and lagoons are too thickly intermingled with the better portion to render it a safe or desirable grazing country. The timber is universally bad and small; a few misshapen gum trees on the immediate banks of the river may be considered an exception.''

The channel of the river now divided, and Oxley followed the channel on the northern side, which they were skirting. But before they had progressed a mile beyond the point of divergence, they reached the spot where the river overflowed its banks and its course was lost in the marshes. It was on the 12th of May that they received this check to their as yet uninterrupted progress.

"Observing an eminence about half-a-mile from the south side, we crossed over the horses and baggage at a place where the water was level with the banks, and which, when within its usual channel, did not exceed thirty or forty feet in width.

"We ascended the hill, and had the mortification to perceive that the termination of our research was reached, at least down this branch of the river. The whole country from the west, north-west, round to the north, was either a complete marsh or lay under water.''

The country to the south and south-west appearing more elevated, Oxley determined to return to the place where the branches separated, and to try his fortune on the other one. This, after a while, proved as unsatisfactory as the one they had abandoned. Bitterly disappointed, Oxley altered his plans entirely. He resolved to cease trying to follow the river through this water-logged country, and determined to strike out on a direct course to the south coast in the neighbourhood of Cape Northumberland. In this way he hoped to cross any river that these dreary marshes and swamps gave birth to, and that found an outlet into the Southern Ocean, between Spencer's Gulf and Cape Otway.

This resolve was at once carried out. The boats were hauled up and secured together; all unnecessary articles were abandoned to suit the reduced means of transit; and at nine o'clock on May 18th they said farewell to this weary river and started to encounter fresh troubles under another guise. Instead of travelling in a superfluity of water they now found themselves straitened by drought, and the work began to tell upon the horses. Scrub, too, that besetting hindrance of so many Australian explorers, began to impede their onward path. Eucalyptus brush overrun with creepers and prickly acacia bushes united to bar the way, and when, after much toil and suffering, they at last reached the point of a range, which Oxley named the Peel Range, the leader had reluctantly again to change his mind and to abandon the idea of making south-west to the coast. Sick at heart at this sequence of disastrous happenings, he confided his feeling of sorrow to his journal.

"June 4th. Weather as usual fine and clear, which is the greatest comfort we enjoy in these deserts abandoned by every living creature capable of getting out of them. I was obliged to send back to our former halting-place for water, a distance of near eight miles; this is terrible for the horses, who are in general extremely reduced; but two in particular cannot, I think, endure this miserable existence much longer.

"At five o'clock two of the men whom I had sent to explore the country to the south-west and see if any water could be found, returned after proceeding six or seven miles; they found it impossible to go any farther in that direction, or even south, from the thick bushes that intersected their course on every side; and no water (nor in fact the least sign of any) was discovered either by them or by those who were sent in search of it nearer our little camp.

"June 5th. From everything I can see of the country to the south-west, it appears, upon the most mature deliberation, highly imprudent to persevere longer in that direction, as the consequences to the horses of want of grass and water might be most serious; and we are well assured that within forty miles on that point the country is the same as before passed over. . . Our horses are unable to go more than eight or ten miles a day, but even they must be assured of finding food of which in these deserts the chances are against the existence."

On the following day, June 6th, Oxley, having changed his course to west and north-west, made another effort to escape from the surroundings that so disheartened him. On the 4th of June, before leaving, Allan Cunningham planted some acorns and peach and apricot stones in honour of the King's birthday. Upon this episode Oxley remarks, that they would serve to commemorate the day and situation, "should these desolate plains be ever again visited by civilised man, of which, however, I think there is very little probability." All this only shows how the lack of experience of the paradoxical nature of the Australian interior induced Oxley to form an absurdly erroneous idea of the country in its virgin state. His observations read almost like a present-day description of the sandy spinifex desert of the north-west of Western Australia, and, in fact, the very same remark was made by Warburton in 1873, when traversing that awful desert. He confessed his uncertainty about the longitude of Joanna Spring, and says that it did not matter, as no white man would ever come into the desert again in search of the oasis.

But Oxley's troubles were increasing, and on June 8th he wrote: "The whole country in these directions, as far as the eye can reach, was one continued thicket of eucalyptus scrub. It was impossible to proceed that way, and our situation was too critical to admit of delay: it was therefore resolved to return back to our last station on the 6th, under Peel's Range, if for no other purpose than that of giving the horses water."

Forced to return once more, Oxley became thoroughly convinced of the inhabitability of the country, and it is no wonder that his condemnation was so sweeping and hasty. He wrote on June the 21st:—

"The farther we proceed westerly, the more convinced I am that for all the practical purposes of civilised man the interior of this country westward of a certain meridian is uninhabitable, deprived as it is of wood, water and grass."

Unfortunately for his fame, he then relinquished all thoughts and hopes of a southward course; for had he pushed on, posterity would have hailed his memory as the discoverer of the Murrumbidgee. But Fate decided otherwise, and dejected and baffled, he turned to follow the Peel Range north, making for the part he had left, where at least he was sure of a supply of water. The expedition suddenly came upon the river again on the 23rd of June, and hoping to find that it had modified its nature, they commenced to run it down again. The 7th of July they were forced to halt once more, when Oxley gave up all idea of tracing the Lachlan. He began his return journey, making this last desponding entry:—

"It is with infinite regret and pain that I was forced to come to the conclusion that the interior of this vast country is a marsh and uninhabitable. . . There is a dreary uniformity in the barren desolateness of this country which wearies one more than I am able to express. One tree, one soil, one water, and one description of bird, fish, or animal prevails alike for ten miles and for one hundred. A variety of wretchedness is at all times preferable to one unvarying cause of pain or distress."

On the 4th of August, the leader, knowing the repellant nature of the river and its swamps and morasses that lay ahead of their returning footsteps, determined to quit the Lachlan altogether, and steering a northern course, to abandon the low country, reach the Macquarie River and follow it up to the settlement at Bathurst.

The boats having been long since abandoned, it was necessary to build a raft of pine-logs wherewith to transport the baggage over the stream. They crossed in safety, and we can imagine that it was with no feelings of regret that they finally lost sight of the stream that had so persistently baffled them in all their attempts to traverse its banks.

For some days they had to struggle against the many obstacles of a new and untrodden land, but they at last emerged on to the Macquarie country, which made a pleasant and welcome contrast with the detested Lachlan.

It may be thought that too much stress has been laid upon Oxley's opinion of the Lachlan, but it was this pessimistic report that dominated the public mind

Photo by Rev. J. M. Curran.

The Lachlan River at the point where Oxley left it on the 4th August, 1818, and struck North-East to gain the Macquarie River and follow that river up to Bathurst.

for many years in its speculations as to the character of the interior.

To Oxley himself, the first glimpse of the Macquarie came like a ray of sunshine on his harassed feelings. Was he not to reap some reward for his heroic efforts along the Lachlan, to enjoy the realisation of some of his ambition as geographical discoverer? The Macquarie seemed a favourable subject for the exercise of his talents. Would it not lead him westward to the conquest of that mysterious inland country which had hitherto guarded its secrets with an invincible obstinacy? Poor Oxley, who can help rejoicing with him in his short-lived joy? Without knowing it, he was the first of a long line of brave spirits who were doomed to lose health and life in carving their way into the heart of Australia.

As they returned homeward up the bank of the Macquarie, the river seemed to him to glitter with the bright promise of a crown of success. For almost the first time the entry in his journal has a cheery ring of hope:—

"Nothing can afford a stronger contrast than the two rivers—Lachlan and Macquarie—different in their habits, their appearance, and the source from which they derive their waters, but, above all, differing in the country bordering on them; the one constantly receiving great accession of water from four streams, and as liberally rendering fertile a great extent of country, whilst the other, from its source to its termination is constantly diffusing and diminishing the water it originally receives over low and barren deserts, creating only wet flats and uninhabitable morasses, and during its protracted and sinuous course is never indebted to a single tributary stream."

(iii.)—The Liverpool Plains.

The disappointment occasioned by Oxley's return to Bathurst and his failure to trace the course of the Lachlan was in part atoned for by the high opinion he had formed of the Macquarie. A second expedition was planned, and the command again offered to the Surveyor-General.

Evans was again second, and Dr. Harris, a very able man, accompanied the party as a volunteer. Charles Frazer was botanist, but Allan Cunningham did not go. The expedition was on a slightly larger scale, there being, besides those already mentioned, twelve ordinary members, with eighteen horses and provisions for twenty-four weeks. A depot was formed at Wellington Valley, and men sent ahead to build two boats.

On June 6th, the start was made from the depot, and for the first 125 miles no obstacles nor impediments were met with. Elated by this, Oxley sent two men back to Bathurst, in accordance with instructions, bearing a favourable despatch to Governor Macquarie. But Fate was again deriding the unfortunate explorer. No sooner had the two parties separated, one with well-grounded hopes of their ultimate success, the other bearing back tidings of these confident hopes, than doubt and distrust entered into the mind of the leader. Twenty-four hours after the departure of the messengers, Oxley wrote in his journal:—

"For four or five miles there was no material change in the general appearance of the country from what it had been on the preceding days, but for the last six miles the land was considerably lower, interspersed with plains clear of timber and dry. On the banks it was still lower, and in many places it was evident that the river-floods swept over them, although this did not appear to be universally the case. . . These unfavourable appearances threw a damp upon our hopes, and we feared that our anticipations had been too sanguine."

And still, as Oxley went on, he found the country getting flatter and more liable to inundation, until at last, with a heart nearly as low as the country, he found himself almost hemmed in by water. In fact, it was necessary to retrace steps in order to find a place where they could encamp with safety. Upon this emergency, Oxley held a consultation with Evans and Harris, and it was decided to send the baggage and horses back to a small and safe elevation that stood some fifteen miles higher up the river, thus making a subsidiary depot camp. Oxley himself, with four volunteers in the largest of the two boats, would take a month's provisions and follow the stream as long as there was enough water to float their craft. Meanwhile, Evans, during Oxley's absence, was to make an excursion to the north-east, and return by a more northerly route, this being the direction the party intended to take, should the river fail them as the Lachlan had done on the previous journey.

It was a wet and stormy day on which Oxley started on the river voyage. For about twenty miles there was, as Oxley expresses it, "no country." The main channels being in an overflow state, the flat country which surrounded them could be recognised only by the timber growing on the banks. The clear spaces whereon no trees grew were now covered with reeds, which stood at the height of six or seven feet above the surface. That night they took refuge on a piece of land which was so nearly submerged that there was scarcely enough space on which to kindle a fire. In the morning the violence of the storm had somewhat abated, and as soon as the grey light was strong enough for them to recognise their way, they resumed their dreary journey.

Oxley still contrived to keep to what he took to be the main channel, although, as it now pursued its course amid a dense thicket of reeds, it was becoming more difficult with every succeeding mile. Oxley's seamanship, however, stood him in good stead, and although fallen logs now began to obstruct their passage, they kept doggedly on for another twenty miles. There was no diminution in the volume of the current that was now bearing them onward, and Oxley felt confident that he was approaching that hidden lake, wherein the inland waters mingled their streams, and of whose existence he thought he had now every reason to rest assured. Just as he was buoying his spirits up with these hopes, dreaming that in future he would be able proudly to say,

> We were the first that ever burst
> Into that silent sea,

the river eluded all further pursuit by spreading out in every direction amongst the ocean of reeds that surrounded them.

Wounded to the heart at this unlooked-for disappointment, Oxley, after vainly seeking for some clue or indication by which he could continue the search, had to 'bout ship and return to the camp of the night before. He says:—

"There was no channel whatever amongst these reeds, and the depth varied from five to three feet."

Although he was still convinced that the "long sought for Australian Sea" existed, he recognised the futility of continuing this search to the westward, in which direction some malignant genius seemed ever to persist in thwarting him; and so he regained the shelter of the depot at Mt. Harris, with another tale of frustrated hopes.

Evans, on his return from his scouting expedition to the north and north-east, had a more cheerful story to tell. The weather had been wet throughout, and the impassable nature of the country occasioned thereby had hampered him greatly; nevertheless he had struggled across the worst of the flat country, and in the north-east had come to a new river, which he named the Castlereagh. He was absent ten days, and on his return Oxley determined to abandon the Macquarie, which had proved even more deceptive and elusive than the Lachlan, and to strike out for the higher lands which Evans reported having seen.

He left Mt. Harris on July 20th, first burying a bottle there containing a written scheme of his intended movements, and some silver coin. Ten years afterwards, Captain Sturt made an ineffectual search for this bottle. Oxley had also buried a bottle at the point of his departure from the Lachlan. Mitchell searched for it without success, and learned afterwards that it had been broken by the blacks.

On July 27th, the party reached the bank of the Castlereagh, after fighting their way through bog, quagmire, and all the difficulties common to virgin country during continued wet weather. As the direction they were steering was towards a range seen by Evans, and named Arbuthnot Range, their march was again interrupted by finding the new-found river this time running bank-high, having evidently risen immediately after Evans had crossed it on his return journey. Here, perforce, they had to stay until the water subsided, and it was not until August 2nd that the river had fallen enough to allow them to cross. The ground was still soaked and boggy, and the horses having had to carry increased pack-loads since the abandonment of the boats, the party suffered great toil and hardship in their efforts to gain Arbuthnot Range. The Range was reached, however, and rounding one end of it by skirting the base of a prominent hill which they named Mt. Exmouth, the harassed explorers at last emerged upon splendid pastoral country.

As Oxley, from a commanding position, surveyed the magnificent scene spread out beneath him—gentle hills separating smiling valleys, which in their turn merged into undulating plains all ripe for settlement—he must have felt that Fate had at length relented, and granted him a measure of reward as the discoverer of this beautiful land. He called the locality Liverpool Plains, and the name has long been synonymous with pastoral prosperity. Their journey to the eastward, which carried them through the heart of this rich and highly-favoured country, was now less arduous; and though the ground was still wet from the late soaking rains, the sun shone cheerily overhead, and the horses, revelling in the abundant rich grass and succulent herbage, began to recover their strength. On September 2nd, they came to a river, which Oxley named the Peel; and here the expedition narrowly escaped the shadow of a fatality, one man being nearly drowned whilst crossing. After leaving the Peel, Oxley still continued easterly, traversing splendid open grazing country. He was now approaching the dividing water-shed of the Main Range, to the northward of that portion of it which is known at the present day as the Liverpool Range. Here the deep glens and gullies with which the seaward front is serrated, began to interfere seriously with the direct course of travel, and at the heads of many of them there were cataracts and waterfalls which compelled the wanderers to turn away to the south; and on one occasion to revert almost to the west. One of these striking natural features received the name of Becket's Cataract, and another was christened Bathurst's Falls. Once again tempests and storms beset them, and this wild weather found them wandering amongst the steep ravines and dizzy descents of the mountainous range, seeking a way leading to the lowlands.

It was on September 23rd that Oxley and Evans, while searching for a practicable route, climbed a tall peak, and from the summit caught a glimpse of the sea. It seems to have greatly impressed Oxley, and he writes in his journal of his emotions on the occasion:—

"Bilboa's ecstacy at the first sight of the South Sea could not have been greater than ours when, on gaining the summit of this mountain, we beheld Old Ocean at our feet. It inspired us with new life; every difficulty vanished, and in imagination we were already home."

The descent was attended with many perils: Oxley says that at one period he would willingly have compromised for the loss of one-third of the horses to ensure the safety of the remainder. But the men with him were tried and steady, and the thick tufts of grass and the loose soil afforded them help in securing a surer footing, of every chance of availing themselves of which the men skilfully took advantage, so that both men and horses reached the foot of the mountain—now called Mt. Seaview—without mishap.

They had reached the head of a river running into the Pacific, and proceeded to follow its course down with more or less difficulty until they reached the mouth, when Oxley, judging the entrance to be navigable, named it Port Macquarie, though one should imagine that he had become tired of that name. The river was named the Hastings.

On October 12th, a toilsome march commenced, following the shore to the southward. The wearisome interruptions of the many inlets and saltwater creeks greatly fatigued and distressed his men. But at last they came upon a boat, half-buried in the sand, which had been lost some time before from a Hawkesbury coaster. This they cleaned and patched, and carried with them, utilising it during the latter stages of this weary journey to facilitate the passage of the many saltwater creeks and channels that impeded their progress. It is owing to the possession of this derelict boat that Oxley crossed the mouth of the Manning without identifying it as a river. The blacks now harassed them greatly, and it was during one of the attacks made upon the party that one of the men, named William Black, was dangerously wounded, being speared through the back and the lower part of the body. The care and conveyance of this invalided man was now added to Oxley's other anxieties, and it was with feelings of great satisfaction that on November 1st they caught sight of the rude buildings of Port Stephens. Through much hardship and privations he had brought his party back without loss.

Oxley sent Evans on to Newcastle with despatches to the Governor, in which he alluded to his sanguine anticipations at the time he had sent in his last report, and their almost immediate collapse. But the discovery of Liverpool Plains compensated in some degree for the disappointment caused by the renewed failure that had attended Oxley's efforts to trace an inland river.

In the following year, 1819, the "Lady Nelson," with the Surveyor-General on board, visited the newly-found Port Macquarie and the Hastings River, to survey the entrance; in which task he was assisted by Lieutenant P. P. King in the "Mermaid." On his return to Port Jackson, in the same year, he made a short excursion to Jarvis Bay with Surveyor Meehan, when they were accompanied by the explorer who was to win fame as

Hamilton Hume. Oxley returned by boat, his companions overland.

(iv.)—The Brisbane River.

It was in October, 1823, that Oxley left Sydney on the expedition that resulted in the finding of the Brisbane River, and the foundation of the settlement at Moreton Bay. He was despatched on a mission to examine certain openings on the east coast, and report on the suitability of them as sites for penal establishments. Moreton Bay, Port Curtis, and Port Bowen were selected; and Oxley left in the colonial cutter "Mermaid," with Uniacke and Stirling as assistants.

As the cutter went up the coast, she called at Port Macquarie, and Oxley had the pleasure of noting the rapid growth of the settlement that had been built upon his recommendation. Further along the coast, Oxley discovered and named the Tweed River. The "Mermaid" reached Port Curtis on the 6th of November, and cast anchor for some time, during which Oxley made a careful examination of the locality, his opinion of it as a site for a settlement being decidedly unfavourable. He however discovered and named the Boyne River.

It being considered too late in the season to proceed and examine Port Bowen, the "Mermaid" went south again, and entering Moreton Bay, anchored off the river that appeared to Flinders to take its source in the Glass House Peaks, and which he had called the Pumice Stone River.

They had scarcely anchored when several natives were seen at a distance, evidently attracted by their arrival, and on examining them with the telescope, Uniacke was struck with the appearance of one of a much lighter colour than that of his companions. The next day Oxley landed and discovered that the man they had noticed was in reality a castaway white man of the name of Pamphlet. He told a singular tale.

He had left Sydney in an open boat with three others, intending to go to the Five Islands and bring back cedar.

A terrible gale arose, and they were blown out to sea and quite out of their reckoning, Pamphlet being under the impression that they had come ashore south of Port Jackson. They had suffered fearful hardships in the open boat, being at one time, he averred, twenty-one days without water, during which time one man died of thirst. The boat was at last cast up on an island in the bay (Moreton Island) where they had joined the blacks, and lived amongst them ever since, a matter of seven months. The other survivors were named Finnegan and Parsons. Pamphlet informed Oxley that not long before the "Mermaid" arrived, the three of them had started to try and reach Sydney overland, but when they had got about fifty miles, he had turned back and the next day had been rejoined by Finnegan, who stated that he had quarrelled with Parsons. The latter was never heard of again.

Finnegan put in an appearance the next day, and Oxley naturally took the opportunity to question them as to the knowledge they had gained of the surrounding country, during their enforced stay in it. On one important point both of them were confident, and this was that, in the southern portion of the bay, a large river was to be found which appeared navigable, having a strong current.

Taking Finnegan with them, Oxley and Stirling started in the whaleboat the following morning to verify this information. They found the river and pulled up it about fifty miles. Oxley was greatly pleased with such a discovery, and landing, ascended a hill which he named Termination Hill. From the top he obtained a view over a wide extent of country, through which he was able to trace the river for a long distance. Strangely enough, the hasty glimpse he thus caught of a new and untrodden part of Australia seemed to confirm his fixed belief in the final destination of the Lachlan and the Macquarie as an inland sea.

"The nature of the country and a consideration of all the circumstances connected with the appearances of the river, justify me in entertaining a strong belief that the source of the river will not be found in mountainous country, but rather that it flows from some lake, which will prove to be the receptacle of those inland streams crossed by me during an expedition of discovery in 1818.''

Oxley named the river the Brisbane, and, taking aboard the two rescued men, the "Mermaid" set sail for Port Jackson, where she arrived on December 13th. This ended the chapter of Oxley's discoveries in the field of active exploration.

Chapter IV.

HAMILTON HUME.

(i.)—Early Achievements.

Hamilton Hume was the son of the Rev. Andrew Hume, who came to the colony with his wife in the transport "Lady Juliana," and held an appointment in the Commissariat Department. Hamilton was born in Parramatta in the year 1797, on the 18th of June. He seems to have been specially marked out by Nature for prominence as an explorer, for, from his earliest boyhood he was fond of rambling through the bush, and his father encouraged him in his desire for a free country life and his love of adventure. School facilities were lacking, but fortunately his mother attended to his education and saw to it that he did not grow up destitute of that instruction common to youth of those times and of his standing.

Hamilton Hume, in his later life.

At the age of seventeen he made his initial effort at exploration in the country around Berrima, in company with his brother Kennedy and a black boy. They were successful in their endeavours, and found some good pastoral country. In the following year, encouraged by their success, the brothers made another excursion. In

1816, a Mr. Throsby bought some of the land that young Kennedy and Hamilton had found; and their father sent them out with him to show him the country he had purchased. John Oxley, too, held a farm in the Illawarra district, and the Surveyor-General, who must have heard of Hamilton's repute for good bushmanship, engaged him to go out with his overseer and guide the men on to the locality. Governor Macquarie also seems to have had his attention drawn to the same conspicuous quality, for he sent young Hume out with Meehan, a surveyor, and Throsby to examine the country about the Shoalhaven River. On the way, however, Throsby disagreed with Meehan about the course they should adopt, and, taking a black boy with him, left his companions and made the best of his way to Port Jervis. Meehan and Hume carried out the work as originally decided on, and then forced their way up the range, which had now seemingly been deprived of a great many of its original terrors by the hardy pioneers of the coast. On the highlands they discovered and named Lake George, a freshwater lake, and a smaller one which they called Lake Bathurst, both, strange to say, seemingly isolated.

Here we may remark on the tenacity with which the Murrumbidgee River long eluded the eye of the white man. It is scarcely probable that Meehan and Hume, who on this occasion were within comparatively easy reach of the head waters, could have seen a new inland river at that time without mentioning the fact, but there is no record traceable anywhere as to the date of its discovery, or the name of its finder. When in 1823 Captain Currie and Major Ovens were led along its bank on to the beautiful Maneroo country by Joseph Wild, the stream was then familiar to the early settlers and called the "Morumbidgee." Even in 1821, when Hume found the Yass Plains, almost on its bank, he makes no special mention of the river. From all this we may deduce the extremely probable fact that the position of the river was shown to some stockrider by a native, who also confided the aboriginal name, and so it gradually worked the knowledge of its identity into general belief. This theory is the more feasible as the river has retained its native name. If a white man of any known position had made the discovery, it would at once have received the name of some person holding official sway. But this is altogether a purely geographical digression.

It was while on this expedition that Hume found the Goulburn Plains. On another occasion he went with Alexander Berry, a noted south-coast pioneer, up the Shoalhaven River, and accompanied the party when they landed and conducted different excursions. By the time he reached manhood, Hume was justly classed amongst the finest bushmen in the colony. In his after career when he led the famous expedition to the south coast, and again, when as Sturt's right hand he accompanied that explorer on the notable expedition that solved the mystery of the outflow of the inland rivers and gave to settled Australia the mighty Darling, he fully proved his right to the title.

(ii.)—Discovery of the Hume or Murray.

It is perhaps by his fame as leader of the party that crossed from Lake George to the Southern Ocean that Hume's name is best remembered. At that time especially it aroused anew the bright hopes for the future of the interior that Oxley's gloomy prognostications had done so much to depress. The Surveyor-General having been unable to determine the question as to whether any large river entered the sea between Cape Otway and Spencer's Gulf, a somewhat hazardous idea entered the head of the then Governor, Sir Thomas Brisbane, to land a party of convicts near Wilson's Promontory, and induce them by the offer of a free pardon and a grant of land to find their way back to Sydney overland. It was further proposed that an experienced bushman should be put in charge of them. The flattering offer of this responsible, if somewhat precarious position, was made to young Hamilton Hume who, on mature consideration, declined it.

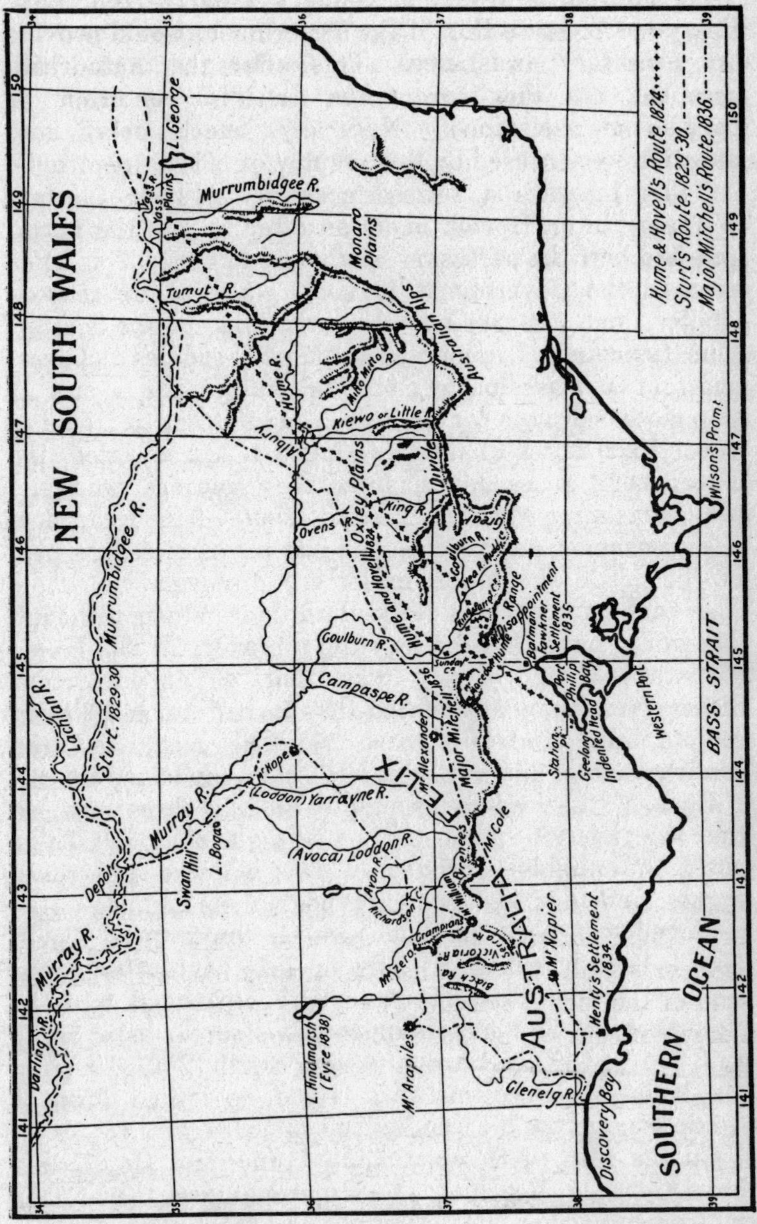
Hume & Hovell's Route, 1824.
Sturt's Route, 1829-30.
Major Mitchell's Route, 1836.
NEW SOUTH WALES
L. George
Yass Plains
Murrumbidgee R.
Tumut R.
Monaro Plains
Australian Alps
Hume R.
Mitta Mitta R.
Kiewo or Little R.
Albury
Oxley Plains
Ovens R.
King R.
Divide
Great
Goulburn R.
Yea R.
Hume and Hovell 1824
Range
Mt Disappointment
Batman Fawkner Settlement 1835
Western Port
Wilson's Promt
BASS STRAIT
Port Phillip Bay
Indented Head
Geelong
Station Pt
Goulburn R.
Campaspe R.
Mt Alexander
Major Mitchell 1836
Mt Hope
(Loddon) Yarrayne R.
Murray R.
Lachlan R.
Sturt. 1829-30
Depot
Swan Hill
L. Boga
(Avoca) Loddon R.
Avon R.
Mt William
Pyrenees
Mt Cole
AUSTRALIA FELIX
Grampians
Mt Zero
Victoria R.
Mt Napier
Henty's Settlement 1834.
SOUTHERN OCEAN
Discovery Bay
Glenelg R.
Mt Arapiles
L. Hindmarsh (Eyre 1838)
Murray R.
Darling R.

He offered, however, to conduct a party from Lake George to Western Port if the Government would provide the necessary assistance. This offer the authorities accepted, but they forgot the essential condition of furnishing assistance. Naturally, much delay and vexation were caused by this display of official ineptitude. At this juncture a retired coasting skipper, Captain William Hilton Hovell, made an offer to join the party, and find half the necessary cattle and horses. This offer aroused the Government to some sense of its responsibility, and it agreed to do something in the matter. This "something" amounted to six pack-saddles and gear, one tent of Parramatta cloth, two tarpaulins, a suit of slop clothes a-piece for the men, and an order to Hume to select 1,200 acres of land for himself. In addition, the Government generously granted the explorers two skeleton charts upon which to trace the route of their journey, a few bushmen, and promised a cash payment for the hire of the cattle should an important discovery be made. This cash payment was refused on their return, although one would have thought that the discovery of the Hume (Murray) should surely take rank as an important discovery. Hume also stated that he had much difficulty in obtaining tickets-of-leave for the men, and the confirmation of his own order to select land for himself.

Each of the leaders brought with him three men, so that the strength of the party was eight all told. Their outfit of animals consisted of five bullocks and three horses, and they had two carts with them.

Hovell was born at Yarmouth on the 26th of April, 1786. He arrived in Sydney in 1813, but after being engaged in the coasting trade with occasional trips to New Zealand, he had relinquished his career as a sailor and had settled at Narellan, New South Wales. After his exploring expedition with Hume, he settled down at Goulburn, and he died at Sydney in 1876.

On the 14th of October, 1824, Hume and Hovell left Lake George. Reaching the Murrumbidgee, they found that river flooded, and after waiting three days for the water to fall, they crossed it borne on the body of one of their carts, with the wheels detached, and with the aid of the tarpaulin, rigged like a punt. South of the Murrumbidgee the country was broken and difficult to traverse, but it was well grassed and admirably adapted for grazing purposes. As it became too rough for the passage of their carts, these were abandoned, and the baggage and rations were packed on the bullocks for the remainder of their journey.

After following the course of the Murrumbidgee for some days, the travellers turned from its bank and pursued a south-westerly direction, which led them through hills and valleys richly grassed and plenteously endowed with running streams. On the 8th of November they beheld a sight rarely witnessed before by white men in Australia. Ascending a range in order to obtain a view of the country ahead of them, they suddenly found themselves confronted with snow-capped mountains. There, under the brilliant sun of an Australian summer's day, rose the white crests of lofty peaks that might have found fitting surroundings amidst the chilling splendours of some far southern clime, robed as they were for nearly one-fourth of their height in glistening snow.

Skirting this range, which received the name of the Australian Alps, the explorers, after wandering for eight days across its many spurs, came upon a fine, flowing river, which Hume named after his father, the Hume. This river was destined to be re-named the Murray, when its course was eventually followed to the ocean.*

There being no safe ford, a makeshift boat was constructed with the aid of the serviceable tarpaulin, and the Hume was crossed, close to the site of the present town of Albury. Still passing through good pastoral land, watered by numerous creeks, they crossed a river which was named the Ovens, and on the 3rd of December they came to another, named by them the Hovell, but now called the Goulburn; and on the 16th of December they

See Chapter VI.

reached their goal, the shore of the Southern Ocean, at the spot where Geelong now stands.

This expedition had a great and immediate influence on the extension of Australian settlement. Within a few years after the chief surveyor had characterised the western interior, beyond a certain limit, as unfitted for human habitation, and had expressed his opinion that the monotonous flats across which he vainly looked for any elevation extended to the sea-coast, snowy mountains, feeding the head tributaries of perennial rivers had been discovered to the southward of his track.

Hume was exceptionally fitted for the work of exploration at this particular juncture in colonial history. Born and reared in the land, he was well competent to judge justly of its merits and demerits; his opinion was not likely to be tainted by the prejudices formed and nourished in other and different climes. The history of Australian exploration was then a statement of hasty conclusions, formed perhaps under certain climatic circumstances to be falsified on a subsequent visit when the conditions were radically different. In Hume's case, there was no ill-founded conclusion of the availability of the freshly-discovered district. The journey just recorded at once added to the British Colonial Empire millions of acres of arable land watered by never-failing rivers, with a climate and altitude calculated to foster the growth of almost every species of temperate fruit or grain.

It is to be regretted that the narration of an expedition fraught with so much benefit to the young colony, and executed with so much courage, endurance, and facility of resource should be marred by any discordant note. But friendly and genial relations were endangered by the presence of two independent leaders. Divided authority here, as it nearly always does, caused petty and undignified squabbles, which were in later days elaborated into unseemly paper conflict. It is painful if somewhat amusing to read of the acrid disputes as to the course, under the very shadow of the majestic Australian Alps whose solitude had only then been first disturbed by white men; and how, on agreeing to separate and divide the outfit, it was proposed to cut the only tent in two, and how the one frying-pan was broken by both men pulling at it. Thomas Boyd, who was the only survivor of the party in 1883, and was then eighty-six years old, signed a document assigning to Hume the full credit of conducting the expedition to safety. Boyd was one of the most active members of the expedition, always to the front when there was any trying work to be done. He was the first white man to cross the Hume River, swimming over with the end of a line in his teeth.

After Hume's return he lived for some time quietly on his farm, until the "call of the wild" drew him forth from his retirement to join Sturt in his first battle with the wilderness. His temporary association with that explorer will find its due place in the account of that expedition.* He died at Yass, near the scene of one of his early exploits.

*See Page 62.

Chapter V.

ALLAN CUNNINGHAM.

(i.)—Coastal Expeditions.

Allan Cunningham, the great botanical explorer of Australia, was born at Wimbledon, near London, in 1791. He received a good education, his father intending him for the law; but he preferred gardening, and obtained a position under Mr. Aiton, at Kew. In 1814 he went to Brazil, where he made large collections of dried specimens, living plants, and seeds. Here he remained two years, collecting in the vicinity of Rio, the Organ Mountains, San Paolo, and other parts of Brazil. Sir Joseph Banks wrote that his collections, especially of orchids, bromeliads, and bulbs, "did credit to the expedition and honour to the Royal Gardens." He was nominated for service in New South Wales, and landed at Port Jackson on the 21st of December, 1816.* He first started collecting about the present suburb of Woolloomooloo in Sydney, which we may infer therefrom presented a very different appearance from that which

Allan Cunningham.

---

*For the accompanying notes of Allan Cunningham's earlier lifework I am indebted to the "Biographical Notes concerning Allan Cunningham," compiled by Mr. J. H. Maiden, Director of the Sydney Botanical Gardens.

it now presents. He next went with Oxley on his Lachlan expedition. On his return, he commenced the first of his five coastal voyages, in which he accompanied Captain P. P. King around most of the continent of Australia. In the tiny cutter the "Mermaid," of 84 tons, they left Port Jackson on the 22nd of December, 1817, and sailed round the south coast of Australia to King George's Sound, the west coast, the north coast, and finally to Timor. The "Mermaid" returned by the same route and anchored in Port Jackson on the 24th of July, 1818. Again on the 24th of December, the "Mermaid" left Port Jackson on a short trip to Tasmania, from which they returned in February, 1819. Once more the busy little "Mermaid" sailed from Sydney on the 8th of May, 1819, to make a running survey of the east coast. On this voyage, many ports hitherto unvisited were examined by King, and amongst other places, Cunningham paid his first visit to the Endeavour River. Continuing the survey, she rounded Cape York, crossed the mouth of the Carpentaria Gulf, and kept along the north coast, where King found Cambridge Gulf. At Cassini Island, the "Mermaid" left for Timor, and eventually returned to Sydney round the west coast of Australia.

On the 14th of June, 1820, the "Mermaid" was again busy with King and Cunningham on board, and, sailing up the east coast she re-visited the Endeavour River. During their stay, Cunningham ascended Mt. Cook, where he made a fine collection of seeds and plants. She coasted north again and picked up the survey at Cassini Island once more. At Careening Bay, where they had occasion to stay some time, Cunningham was again very fortunate in his collections. Returning homeward by way of the west and south coasts, the little cutter was almost wrecked off Botany Bay.

The "Mermaid" was now overhauled and condemned, and in her place H.M. Storeship "Dromedary," rechristened the "Bathurst," was placed under the command of Lieutenant King. This was Cunningham's fifth voyage as collector with the same commander—a very clear proof of their compatibility of tastes and temperament. As before, the "Bathurst" ran round the east coast and resumed her work on the north-west of Australia. While thus engaged she was found to be in a dangerous condition, and went to Port Louis to refit. They sailed from Mauritius on the 15th of November, and reached King George's Sound on the 24th of December. Here Cunningham found that the garden he had been at great pains to form during his visit in 1818 had disappeared altogether. The "Bathurst" stayed some weeks on the south-west coast, and then shaped a course to Port Jackson, where they arrived on the 25th of April, 1822. Of the botany of these coastal surveys Cunningham published a sketch entitled, "A Few General Remarks on the Vegetation of Certain Coasts of Terra Australis, and more especially of its North-Western Shore."

(ii.)—Pandora's Pass.

Let us now turn to his record as an inland explorer of Australia.

On the 31st of March, 1823, Allan Cunningham left Bathurst with two objects in view. One was his favourite pursuit of botany; and the other the discovery of an available route to Oxley's Liverpool Plains, through the range that bounded it on the south; a route which Lawson and Scott had vainly sought for the preceding year. On reaching the vicinity of the range, he searched in vain to the eastward for any opening that would enable him to pierce the barrier. He then retraced his steps, and, exploring more to the eastward, he came upon a pass through a low part of the mountain belt which he considered practicable and easy. The valley leading to the pass he named Hawkesbury Vale, and the pass itself Pandora's Pass, inasmuch as, in spite of the hardships the party had been put to, they had still hoped to find it. Here Cunningham left a parchment document, stating that the information thereon contained was for the first farmer "who may venture to advance as far to the northward as this vale." The finding of the bottle

which contained this scroll, has never been recorded.
Bathurst was reached on their return journey, on June
27th.

In March, 1824, he botanised about the heads of the
Murrumbidgee and the Monaro and Shoalhaven Gullies,
and in September of the same year, went north by sea
with Oxley to Moreton Bay, to investigate that locality
and pronounce on its suitability as a settlement site. In
March, 1825, he left Parramatta, threaded the Pandora
Pass once more, and ascended to Liverpool Plains,
returning to Parramatta on the 17th of June. In 1826
and the beginning of the following year, he visited New
Zealand.

(iii.)—The Darling Downs.

It was in the year 1827 that Cunningham accomplished
his most notable journey of exploration, one which
eventually threw open to settlement an entirely new area
of country; country destined to mould the destiny of the
yet unborn colony of Queensland, and afford homes for
thousands of settlers. It was mainly by his exertions
that the young community at Moreton Bay was able to
stretch its growing limbs to the westward immediately
after its birth, instead of waiting long weary years and
wasting its strength against an impassable obstacle as
had been the fate of the settlement at Farm Cove.

Cunningham started from Segenhoe, a station on one
of the head tributaries of the Hunter River, whence
he ascended the main range without any difficulty beyond
having to unload some of the pack-horses during the
steepest part of the ascent. He had with him six men,
eleven horses, and provisions for fourteen weeks. He
left civilisation, or the outskirts of it, on the 2nd of May,
and on the 11th he crossed the parallel on which Oxley
had crossed the Peel River in 1818, and once beyond that
point he was traversing unexplored country. The land
was suffering under a prolonged drought in that district,
and most of the streams encountered had but detached
pools of water in their beds, at one of which, however,
his party caught a good haul of cod, which were such

ravenous biters and so heavy that several were lost in
the attempt to land them.

Travelling through open forest land, which was
suffering more or less from the want of rain, Cunningham
came on the 19th of May to a valley. Here, on the bank
of a creek he encamped on "the most luxuriant pasture
we had met since we had left the Hunter."

"We were not a little surprised," he says, " to observe
at this valley, so remote from any farming establishment,
the traces of horned cattle, only two or three days old,
as also the spots on which about eight to a dozen of these
animals had reposed. . .

"From what point of the country these cattle had
originally strayed appeared at first difficult to determine.
On consideration, however, it was thought by no means
impossible that they were stragglers from the large wild
herds that are well-known to be occupying plains around
Arbuthnot Range."

This speaks volumes for the wonderful increase and
spread of wild cattle in those days; Arbuthnot Range,
first sighted by Evans in 1817, being already an
acknowledged resort of wild cattle in seven years. Or
it advertises the negligence of the stockmen who guarded
the comparatively tiny herds of the period.

The dry weather had put its mark upon the country.
Though the degree of aridity was much less than that
afterwards experienced in Australia by the explorers
of its interior, nevertheless conditions were sufficiently
dry to compel the leader to exercise great forethought,
and Cunningham determined to pursue a more easterly
course, keeping nearer the crest of the range, where he
was more likely to find grass and water. The country
he passed through was inferior, but on the 28th he came
to the bank of a river "presenting a handsome reach,
half-a-mile in length, thirty yards wide, and evidently
very deep." This river he named the Dumaresque, and
it led him to the northward, through what he considered
poor land, until the new-found river took an easterly
direction, when the party left it, still keeping north. At
the end of the month, after passing through much scrubby

country, they were agreeably surprised to meet with a stream, the banks of which presented an appearance of great verdure. "It was a subject of great astonishment to us to meet with so beautiful a sward of grass permanently watered by an active stream, after traversing that tract of desert forest, and penetrating brushes the extremes of sterility in its immediate vicinity."

This was named McIntyre's Brook, and Cunningham writes that they had some difficulty in fording it on account of its extreme rapidity. The party continued on, now in a north-easterly direction, passing again through dense thickets such as they had formerly met with.

On the 5th of June, Cunningham, from a small elevation, had a view of open country of decidedly favourable appearance:—"A hollow in the forest ridge immediately before us allowed me distinctly to perceive that at a distance of eight or nine miles, open plains or downs of great extent appeared to extend easterly to the base of a lofty range of mountains, lying south and north, distant by estimation about thirty miles."

This was Cunningham's first glimpse of the now world-famous Darling Downs. On reaching the commencement of the great plains, they came to the "bank of a small river, about fifteen yards in breadth, having a brisk current to the N.W." As there was deep water in the pools of this river, the men anticipated some good fishing, and they were not disappointed. Cunningham named this river the Condamine.

Although their provisions were failing them, Cunningham remained for some time on the site of his new discovery, fully impressed with the certainty of its immense importance in the future settlement of Australia. Peel's Plains and Canning Downs were named by him, and to the north-west "beyond Peel's Plains an immeasurable extent of flat country met the eye, on which not the slightest eminence could be observed to interrupt the common level, which, in consequence of the very clear state of the atmosphere, could be discerned to a very distant blue line of horizon.''

Cunningham's far-seeing mind fathomed the future requirements of such a vast agricultural and pastoral extent of country, and he at once turned his attention to its natural means of communication with its obvious port, Moreton Bay. A lofty range of mountains to the east and north-east seemed to offer a difficult barrier, and he determined upon making a closer inspection. As his horses were recruiting all the time on the luxuriant herbage, he did not so much regret their own scarcity of rations. Finding a beautiful grassy valley which he named Logan Vale, after Captain Logan, the well-known commandant of Moreton Bay, leading to the base of the principal range, he proceeded to make a nearer inspection. After much climbing of successive tiers or ridges, he gained the loftiest point of a main spur, and through some gaps in the main range itself, he was able to overlook portions of the country in the vicinity of Moreton Bay, and even to recognise the cone of Mount Warning. He took particular notice of one gap, and on closer inspection he came to the conclusion that a line of road could be constructed without much difficulty.

Having spent a week on the Downs, and his shortness of provisions and the weakness of his horses preventing any excursion to the western interior, as his intention had been, he set out on his homeward journey on the 18th of June. In order to render his chart of the country traversed as complete as possible, he kept a course about equidistant between the route of his outward journey and the coastal watershed. He reached Segenhoe on the 28th of July, bringing his men and horses back in safety, after one of the most successful and important expeditions on the east coast.

In the following year, accompanied by his old companion Fraser, who had been one of Oxley's party on his two inland expeditions, Cunningham proceeded by sea to Moreton Bay, with the intention of starting from the settlement, identifying the gap he had taken particular notice of, and connecting with his former camp

Memorial to Allan Cunningham, Botanical Gardens, Sydney.

on the Downs. In this attempt he was also accompanied
by Captain Logan, but they were unsuccessful. Then
Cunningham again went from the outpost of Limestone,
with three men and two bullocks, and was completely
satisfied. A road through this gap on to the Darling
Downs was immediately constructed, and used until the
introduction of railway communication: the opening was
known far and wide as Cunningham's Gap.

In May, 1830, Cunningham went to Norfolk Island.
While there he crossed to the little islet adjoining, known
as Phillip Island. Having landed with three men, he
sent the boat back. That night eleven convicts escaped,
seized the boat, and were launching her when they were
challenged by a sentry. One of them replied that they
were going for Mr. Cunningham, and they got away
though they were fired upon. They did go for Mr.
Cunningham, and robbed him of his chronometer, pistols,
tent, and provisions. Then they sailed away, and were
picked up by a whaler, which they seized and finally
scuttled. The Government refused to compensate
Cunningham for his loss, and he had to replace the
instruments himself.

Cunningham left Sydney on the 25th of February,
1831, on a visit to London, where he spent nearly two
years at Kew, returning to Sydney on the 12th of
February, 1837. He was appointed Colonial Botanist
and Superintendent of the Botanic Gardens, but did not
retain the position very long, being disgusted to find that
supplying Government officials with vegetables was to be
a chief part of his duties. He resigned, and after another
visit to New Zealand, whence he returned in 1838, so ill
was he that he was compelled to decline to accompany
Captain Wickham on his survey of the north-west coast.
He died of consumption on the 24th of January, 1839,
at the cottage in the Botanic Gardens, whither he had
been removed for change of air and scene. He was
buried in the Devonshire Street cemetery, and on the 25th
of May, 1901, his remains were removed to the obelisk
in the Botanic Gardens.

Chapter VI.

CHARLES STURT.

(i.)—Early Life.

Charles Sturt was born in India at Chunar-Ghur, on April the 28th, 1795. His father, Thomas Lennox Napier Sturt, was a puisne Judge in Bengal under the East India Company; his mother was Jeanette Wilson. The Sturts were an old Dorsetshire family. In 1799, Charles, as was common with most Anglo-Indian children, was sent home to England, to the care of his aunts, Mrs. Wood and Miss Wilson, at Newton Hall, Middlewich. He went first to a private school at Astbury, and in 1810 was sent to Harrow. On the 9th of September, 1813, he was gazetted as Ensign in the 39th Regiment of Foot. He served with his regiment in the Pyrenees, and in a desultory campaign in Canada. When Napoleon escaped from Elba, the 39th returned to Europe, but all too late to join in the victory of Waterloo, and it was stationed with the Army of Occupation in the north of France. In 1818, the regiment was sent to Ireland. Here for several years Sturt remained in most uncongenial surroundings, watching smugglers, seizing illicit stills, and assisting to quell a rising of the Whiteboys. It was in Ireland that the devoted John Harris, his soldier-servant, who was afterwards the companion of his Australian wanderings, was first attached to him. In 1823, Sturt was gazetted Lieutenant, and his promotion to Captain followed in 1825.

In December, 1826, he sailed for New South Wales with a detachment of his regiment, in charge of convicts. The moment he set foot on this vast unknown land, its chief geographical enigma at once occupied his attention.

Sir Ralph Darling, to whom he acted for some time as private secretary, formed a high opinion of his tact and ability, and appointed him Major of Brigade and Military Secretary.

(ii.)—The Darling.

As soon as an expedition inland was mooted, Sturt volunteered for the leadership, and was recommended by Oxley, who was then on his deathbed. The recommendation was adopted by Governor Darling, and Sturt embarked on the career of exploration that was to render his name immortal.

It was ever Sturt's misfortune to be the sport of the seasons; drought and its attendant desolation dogged his footsteps like an evil genius. Oxley had followed, or attempted to follow, the rivers down when a long period of recurrent wet seasons had saturated the soil, filled the swamps and marshes, and swollen the river-courses so that they appeared to be navigable throughout for boats. Sturt came at a period when the country lay faint under a prolonged drought and the rivers had dwindled down into dry channels, with here and there a parched and meagre water-hole. The following description of his is too often quoted as depicting the usual state of the Australian interior:—

"In the creeks, weeds had grown and withered, and grown again; and young saplings were now rising in their beds, nourished by the moisture that still remained; but the large forest trees were drooping, and many were dead. The emus with outstretched necks, gasping for breath, searched the channels of the rivers for water in vain; and the native dog, so thin that he could hardly walk, seemed to implore some merciful hand to despatch him."

To Sturt and his companions, who were the first white men to face the interior during a season of drought, the scene may not have seemed too highly-coloured; but, in common with many of Sturt's graphic word-pictures, his description applies only to special or rare circumstances.

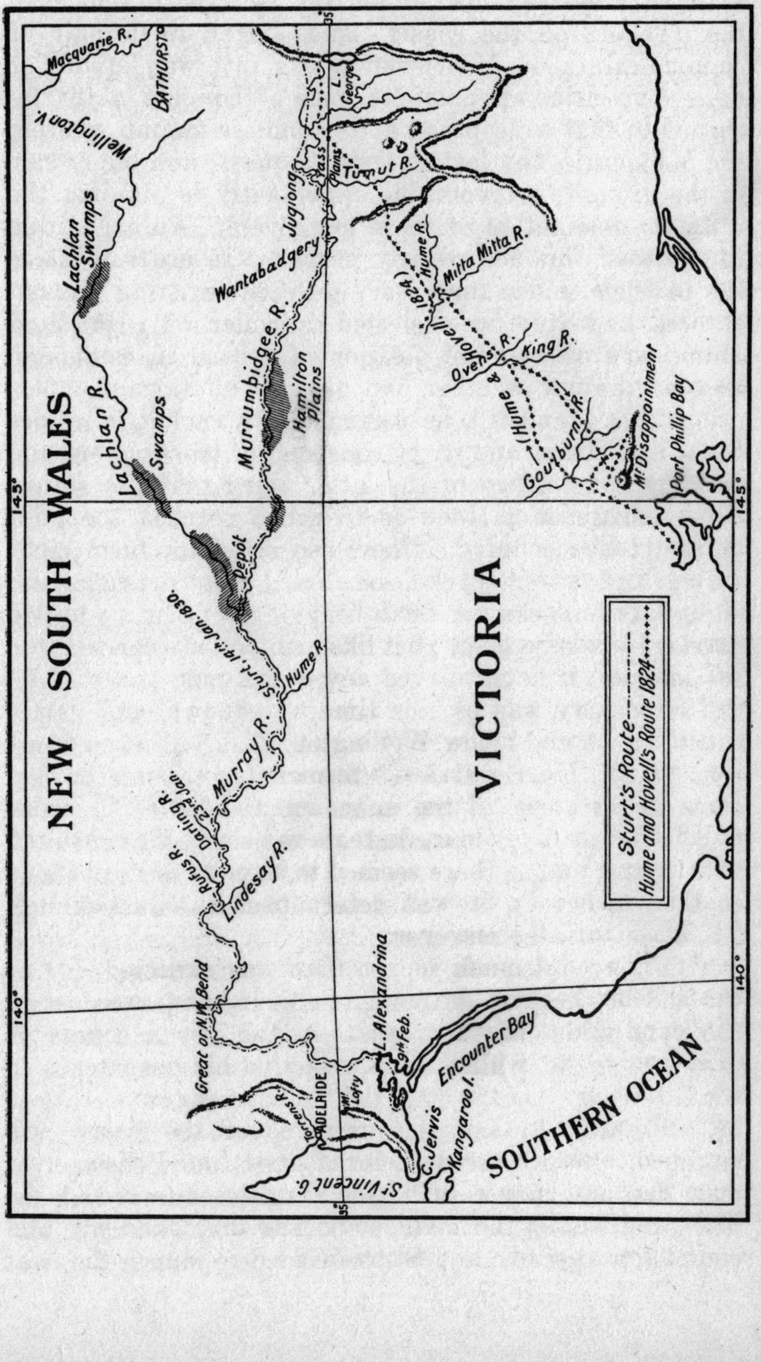
NEW SOUTH WALES
VICTORIA
SOUTHERN OCEAN
Macquarie R.
BATHURST
Wellington V.
Lachlan Swamps
Lachlan R.
Swamps
Depôt
Sturt, 19th Jan. 1830.
Hume R.
Murray R.
Darling R.
23rd Jan.
Rufus R.
Lindesay R.
Great or N.W. Bend
Murrumbidgee R.
Hamilton Plains
Wantabadgery
Jugiong
Yass Plains
L. George
Tumut R.
Hume R.
Mitta Mitta R.
Hume & Hovell 1824
Ovens R.
King R.
(Hume
Goulburn R.
Mt. Disappointment
Port Phillip Bay
L. Alexandrina
9th Feb.
Encounter Bay
Kangaroo I.
C. Jervis
St. Vincent G.
ADELAIDE
Mt. Lofty
Torrens R.
Sturt's Route
Hume and Hovell's Route 1824
35°
140°
145°

In 1828, no rain had fallen for two years, and even the dwellers on the coastal lands began to despair of copious rainfalls. Whenever their glance wandered over their own dried-up pastures, men's thoughts naturally turned to that widespread and boundless swamp wherein the Macquarie was lost to Oxley's quest; and many saw in the drought a favourable opportunity to discover the ultimate destination of these lost rivers. An expedition to the west was accordingly prepared in order to solve the problem under these very different existing circumstances, and Sturt was selected as leader. To Hamilton Hume was offered the position of second in command, and, as the dry weather had brought all farming operations to a standstill, he was able to accept it. Besides Sturt and Hume, the party consisted of two soldiers and eight prisoners, two of the latter being taken to return with despatches as soon as they had reached the limit of the known country. They also had with them eight riding and seven pack-horses, and two draught and eight pack-bullocks. A small boat rigged up on a wheeled carriage was also taken; but like many others carried into the interior, it never served any useful purpose.

The country was by this time well-known, and partly settled up to and below Wellington Vale; but when Sturt reached Mt. Harris, Oxley's former depot camp, he had come to the verge of the unknown, and halted in order to consider as to his immediate movements. He consulted with Hume, and as there seemed to be no present obstacle to their progress, it was determined, as Sturt writes, "to close with the marshes."

This they did much sooner than was expected, for at the end of the first day's march their camp was set in the very midst of the reeds. A halt for a couple of days was made, whilst Sturt prepared his despatches to the Governor. On the 26th, the two messengers were sent off to Bathurst, and the progress of the party was resumed. Before the day closed, they found themselves on a dreary expanse of flats and of desolate reed beds. The progress of the main body was thus suddenly and completely checked, and Sturt decided to launch the boat

and with two men endeavour to trace the course of the
river, while Hume and two others endeavoured to find an
opening to the northward.

The boat voyage soon terminated, for Sturt was as
completely baffled as Oxley had been. The channel ceased
altogether, and the boat quietly grounded. Sturt could
do nothing but return to camp and await Hume's report.
All search for the lost river proved vain.

Hume had found a serpentine sheet of water to the
north which he was inclined to think was the continuation
of the elusive Macquarie. He had pushed on past it,
but had been checked by another body of reed beds.
It was decided to shift camp to this lagoon and launch
the boat once more; but without result, for the boat was
hauled ashore again after having vainly followed the
supposed channel in amongst reeds and shallows. Again
the leader and his second went forward on a scouting
trip. Each took with them two men; Sturt going to
the north-west, and Hume to the north-east. They left
on the last day of December, 1828.

Sturt toiled on until after sunset he came to a
northward-flowing creek, in which there was a fair supply
of water. Next day their course lay through plains intersected with belts of scrub, and they discovered another
creek, inferior to the last one both in size and the quality
of the water. They camped for a few hours on its bank,
and Sturt called it New Year's Creek, but it is now known
as the Bogan River. They were about to pass that night
without water on the edge of a dry plain, when one of
the men had his attention drawn to the flight of a
pigeon, and searching, found a puddle of rain water
which barely satisfied them. An isolated hill with perpendicular sides, which Sturt had noticed for some time,
now attracted his attention, as being a lofty point of
vantage from which to get an extensive view to the west.
They accordingly made for it, over more promising
country. They reached the hill which Sturt called Oxley's
Tableland, but from its summit he saw nothing but a
stretch of monotonous plain, with no sign of the long-
sought river. That night they camped at a small swamp,

and the next morning turned back, Sturt agreeing with
Oxley, but without as much reason, that "the space I
traversed is unlikely to become the haunt of civilised
man." Hume did not return until the day after Sturt's
arrival. He reported that the Castlereagh River must
have suddenly turned to the north below where Oxley
crossed it, for he had been unable to find it. He had
gone westward, but had seen nothing except far-
stretching plains. After a few aimless and unprofitable
ramblings, they made their way again to Oxley's Tableland, and Sturt and Hume, with two men, made a journey
to the west, with only a negative result. On the 31st of
January they commenced to follow down Sturt's New
Year's Creek, and the next day, to their unbounded
surprise, came upon the bank of a noble river. From its
size and width they judged they had struck it at a point
as far from its source as from its termination; but when
the men rushed tumultuously down the bank to revel in
the water and quench their thirst, they cried out, with
disgust and surprise, that the water was salt.

Poor Sturt, whose heart was bounding with joy at the
realisation of his fondest hopes in this important
discovery of a river which seemed to answer all men's
dreams and anticipations, felt the sudden revulsion of
despair. One saving thought he had, and that was that
they were close to its junction with the inland sea. Meantime, although human tracks were to be seen everywhere,
they saw none of the aborigines. Hume at length found
a pool of fresh water, which provided them with water
for themselves and their stock.

The long-continued absence of rain having lowered the
fresh water so that the supply from the brine springs
on the banks predominated, was the explanation of the
saltness of the water; but Sturt did not know this, and for
six days the party moved slowly down the river until
the discovery of saline springs in the bank convinced
the leader that the saltness was of local origin. Still
that did not supply them with the necessary drinking
water, and on the sixth day, leaving the men encamped
at a small supply of fresh water, Sturt and Hume pushed

Photo by Rev. J. Milne Curran.

The Darling River, at Sturt's first view point.

on to look for more, but in vain, and Sturt was compelled to order a retreat to Mt. Harris.

This shows how the exploration of the continent has ever been conditioned by the uncertainty of the seasons. Had Sturt found the Darling in a normal season, he would probably have followed it down to its junction with the Murray, and the geographical system of the east would have been at once laid bare. But it was not in such a simple manner that the great river basin was to become known. Toil, privation, and the sacrifice of human lives, had first to be suffered.

To the river which he had found Sturt gave the name Darling, in honour of the Governor.

The return journey to Mt. Harris continued without interruption. At Mt. Harris they expected to find fresh supplies; but as they approached the place they could not restrain fears with regard to their safety. The surrounding reed beds were in flames in all parts. The few natives that were met with displayed a guilty timidity, and one was observed wearing a jacket. Fortunately, however, their fears were groundless; the relief party had arrived and had been awaiting their return for about three weeks. An attack by the natives had been made, but it had been easily repulsed. While Sturt rested at Mt. Harris, Hume struck off to the west, beyond the reeds. He reported the country as superior for thirty miles to any they had yet seen, but beyond that limit lay brushwood and monotonous plains.

On the 7th of March the party struck camp and departed for the Castlereagh River. They found that the flooded stream, impassable by Oxley, had totally disappeared. Not a drop of water lay in the bed of the river. They commenced to follow its course down, and the old harassing hunt for water had to be conducted anew. No wonder that Sturt could never free himself from the memory of his fiery baptism as Australian explorer, and that his mental picture of the country was ever shrouded in the haze of drought and heat.

As they descended the Castlereagh into the level lower country, they were greatly delayed by the many intricate windings of the river and its multiplicity of channels. On the 29th of March they again reached the Darling, ninety miles above the place where they had first come upon it, and they observed the same characteristics as before, including the saltness. This was a blow to Sturt, who had hoped to find it free from salinity. Fortunately they were not distressed for fresh water at the time, and knowing what to expect if the river was followed down again, the party halted and formed a camp.

The next day Sturt, Hume, and two men crossed the river and made a short journey of investigation to the west, to see what fortune held for them further afield. Not having passed during the day "a drop of water or a blade of grass," they found themselves by mid-afternoon on a wide plain that stretched far away to the horizon. Sturt writes that had there been the slightest encouragement afforded by any change in the country, he would even then have pushed forward, "but we had left all traces of the natives behind us, and this seemed a desert they never entered—that not even a bird inhabited."

Back to Mt. Harris once more, where they arrived on the 7th of April, 1829. On their way they had stopped to follow a depression first noticed by Hume, and decided that it was the channel of the overflow of the Macquarie Marshes.

(iii.)—The Passage of the Murray.

The mystery of the Macquarie was now, to a certain extent, cleared away, but the course and final outlet of the Darling now presented another riddle, which Sturt too was destined to solve.

The discovery of such a large river as the Darling, augmented by the Macquarie and Castlereagh, and (so people then thought) in all probability the Lachlan, naturally inflamed public curiosity as to the position of the outlet on the Australian coast. All the rivers that had been tried as guides to the hidden interior having failed to answer the purpose, the Murrumbidgee—the beautiful river of the aboriginals—was selected as the scene of the next attempt. There were good reasons for the choice: it derived its volume from the highest known mountains, snow-capped peaks in fact, that reminded the spectator of far northern latitudes, and thus it was to a great extent independent of the variable local rainfall.

Captain Sturt was naturally selected to be the leader of the Murrumbidgee expedition, and with him as second went George MacLeay, the son of the then Colonial Secretary. Harris, who had been Sturt's soldier-servant for nearly eighteen years, and two other men of the 39th, who had been with their Captain on the Macquarie expedition, also accompanied him, with a very complete and well-furnished party, including the usual boat rigged up on a carriage. This time, however, unlike the craft that had accompanied previous exploring parties, the whale-boat was destined to be immortalised in Australian history.

Settlement had by this time extended well up to and down the banks of the Murrumbidgee, and Sturt took his departure from the borders of civilisation about where the town of Gundagai now stands, almost at the junction of the Tumut River, at Whaby's station. The course for some time lay along the rich river-flats of the Murrumbidgee. The blacks, who of course from their position were familiar with the presence of white men, maintained a friendly demeanour. One slight excursion to the north was made to connect with Oxley's furthest south, made when on his Lachlan expedition; but though they did not actually verify the spot, Sturt reckoned that he went within twenty miles of it, showing how narrowly that explorer had missed the discovery of the Murrumbidgee.

As they got lower down the river they found themselves travelling through the flat desolate country that reminded them only too forcibly of late experiences on the Macquarie. Owing to some information gleaned from the natives, Sturt and MacLeay rode north to try and again come upon the Lachlan. They struck a dry channel, which Sturt believed was the drainage from the Lachlan into the Murrumbidgee. This proved to be correct, as natives afterwards testified that they had seen the two white men actually on the Lachlan.

On the 25th, which was an intensely hot day, MacLeay, who was on ahead, found himself suddenly confronted with a boundless sea of reeds, and the river itself had suddenly vanished. He sent a mounted messenger back to Sturt with these disastrous tidings. Sturt thereupon turned the drays, which were already in difficulties in the loose soil, sharp round to the right, and finally came to the river again, where they camped to discuss the untoward circumstance.

At daylight the next morning, Sturt and MacLeay rode along its bank, whilst Clayton, the carpenter, was set to work felling a tree and digging a sawpit. Progress along the bank with the whole party was evidently impossible. Sturt, however, had faith in the continuity of the river, and announced to MacLeay his intention to send back most of the expedition, and with a picked crew to embark in the whaleboat, committing their desperate fortunes to the stream, and trusting to make the coast somewhere, and leaving their return in the hands of Providence.

The more one regards this heroic venture, the more sublime does it appear. The whole of the interior was then a sealed book, and the river, for aught Sturt knew, might flow throughout the length of the continent. But the voyage was commenced with cool and calm confidence.

In a week the whaleboat was put together, and a small skiff also built. Six hands were selected for the crew, and the remainder, after waiting one week in case of accident, were to return to Goulburn Plains and there await events. It would be as well to embody here the names of this band. John Harris, Hopkinson, and Fraser were the soldiers chosen, and Clayton, Mulholland, and Macmanee the prisoners. The start was made at seven on the morning of January 7th, the whale-boat towing the small skiff. Within about fifteen miles of the point of embarkation they passed the junction of the Lachlan, and that night camped amongst a thicket of reeds. The next day the skiff fouled a log and sank, and though it was raised to the surface and most of the contents recovered, the bulk of them was much damaged. Fallen and sunken logs greatly endangered their progress, but on the 14th they "were hurried into a broad and noble river." Such was the force with which they were shot out of the Murrumbidgee that they were carried nearly to the opposite bank of the new and ample stream. Sturt's feelings at that moment were to be envied, and for once in a life chequered with much disappointment he must have felt that a great reward was granted to him in this crowning discovery. He named the new river the Murray, after Sir George Murray, the head of the Colonial Department. As some controversy has of late arisen as to the question of Sturt's right to confer the name, we here quote his own words, written after surveying the Hume in 1838.

"When I named the Murray I was in a great measure ignorant of the other rivers with which it is connected. . . . I want not to usurp an inch of ground or of water over which I have not passed."

On the bosom of the Murray they could now make use of their sail, which the contracted space in the bed of the Murrumbidgee had before prevented them from doing. The aborigines were seen nearly every day, and once when the voyagers had to negotiate a very ticklish rapid, some of them approached quite close, and seemed to take great interest in the proceedings.

Sturt's thoughts now turned towards the junction of the Darling, and at last he sighted a deserted camp on which the huts resembled those he had seen on that river. On the 23rd of January they came upon the junction at a very critical moment. A line of magnificently-foliaged trees came into view, among which was perceived a large gathering of blacks, who apparently were inclined to be hostile. Sturt, who was at the helm, was steering straight for them and made the customary signs of peace. Just before it was too late to avoid a collision, Sturt marked hostility in their quivering limbs and battle-lusting eyes. He instantly put the helm a-starboard, and the boat sheered down the reach, the baffled natives running and yelling defiantly along the bank. The river, however, was shoaling rapidly, and from the opposite side there projected a sand-spit; on each side of this narrow passage infuriated blacks had gathered, and there was no mistaking their intentions. Sturt gave orders to his men as to their behaviour, and held himself ready to give the battle-signal by shooting the most active and forward of their adversaries.

Mention has been made of a small party of blacks who had been interested in the shooting of a rapid by the boat's crew. Four of these savages had camped with the explorers the preceding night, leaving at daylight in the morning. Sturt imagined that they had gone ahead as

Junction of the Darling and Murray Rivers.

peace delegates, and he was thus most anxious to avoid a fight. But the life of the whole party depended on prompt action being taken, and Sturt's eye was on the leader and his finger on the trigger when "my purpose," he says "was checked by MacLeay, who called to me that another party of blacks had made their appearance on the left bank of the river. Turning round, I observed four men at the top of their speed." These were the dusky delegates, and the description given by Sturt of the conduct of the man who saved the situation is very graphic:—

"The foremost of them, as soon as he got ahead of the boat, threw himself from a considerable height into the water. He struggled across the channel to the sand-bank, and in an incredibly short space of time stood in front of the savage against whom my aim had been directed. Seizing him by the throat, he pushed him backwards, and forcing all who were in the water on the bank, he trod its margin with a vehemence and an agitation that was exceedingly striking. At one moment pointing to the boat, at another shaking his clenched hand in the faces of the most forward, and stamping with passion on the sand, his voice, that was at first distinct, was lost in hoarse murmurs.''

This episode, unequalled in the traditions of the Australian aborigines, removed the imminent danger; and Sturt's tact, in a few moments changed the hundreds of demented demons into a pack of laughing, curious children, an easy and common transition with the savage nature. But for the intervention of this noble chief, Sturt and his followers, penned within the boat in shallow water, would have been massacred without a chance to defend themselves. Surrounded as they were by six hundred stalwart foes, their fate, save from unreliable native tradition, would never have been known to their countrymen.

During the crisis, the boat had drifted untended, and grounded on the sand. While the men were hastily pushing her off, they caught sight of "a new and beautiful stream coming apparently from the north." A crowd of natives were assembled on the bank of the new river, and Sturt pulled across to them, thus creating a diversion amongst his erstwhile foes, who swam after, as he says, "like a parcel of seals."

After presenting the friendly native with some acknowledgment and refusing presents to the others, the pioneers examined the new river. The banks were sloping and well-grassed, crowned with fine trees, and the men cried out that they had got on to an English river. To Sturt himself the moment was supreme. He was convinced "that we were now sailing on the bosom of that very stream from whose banks I had been twice forced to retire." They did not pull far up the stream, for a native fishing-net was stretched across, and Sturt forbore to break it. The Union Jack was, however, run up to the peak and saluted with three cheers, and then with a favouring wind they bade farewell to the Darling and the now wonderstruck natives.

As they went on, the party landed occasionally to inspect the surrounding country, but on all sides from their low elevation they could see nothing but a boundless flat. The skiff being now only a drag upon them, it was broken up and burnt for the sake of the ironwork. On account of the damage to the salt pork caused by the sinking of this boat, the strictest economy of diet had to be exercised, and though an abundance of fish was caught, they had become unattractive to their palates. The continuation of the voyage down the course of the Murray was henceforth a monotonous repetition of severe daily toil at the oar. The natives whom they encountered, though friendly, became a nuisance from the constant handling and embracing that the voyagers had, from purposes of policy, to suffer unchecked. The tribes met with were more than ordinarily filthy, and were disfigured by loathsome skin diseases. After twenty-one days on the water, Sturt began to look most anxiously for indications of the sea, for his men were fagging with the unremitting labour and short rations, and they had only the strength of their own arms to rely on for their return against the current. Soon, however, an old man amongst the natives described the roaring of the waves, and showed by other signs that he had been to the sea coast. But more welcome than all were some flocks of sea-gulls that flew over and welcomed the tired men.

On the thirty-third day after leaving the starting-point on the Murrumbidgee, Sturt, on landing to inspect the country, saw before him the lake which was indeed the termination of the Murray, but not the end that he had dreamt of. "For the lake was evidently so little influenced by tides that I saw at once our probable disappointment of practical communication between it and the ocean."

This foreboding was realised after examination of Lake Alexandrina, as it is now called. Upon ascertaining their exact position on the southern coast, nothing was left but to take up the weary labours of their return; the thunder of the surf brought no hopeful message of succour, but rather warned the lonely men to hasten back while yet some strength remained to them.

Sturt re-entered the Murray on his homeward journey on the 13th of February; and the successful accomplishment of this return is Sturt's greatest achievement. His crew were indeed picked men, but what other Australian leader of exploration could have inspired them with such a deep sense of devotion as to carry them through their herculean task without one word of insubordination or reproach. "I must tell the Captain to-morrow that I can pull no more," was the utmost that Sturt heard once, when they thought him asleep; but when the morrow came the speaker stubbornly pulled on.

Three of these men, it must be remembered, were convicts; yet, despite their heroic conduct, one only (Clayton) received a free pardon on their return, though Sturt did his utmost to win fuller recognition of their merits.

In such a work of generalisation as this, space will not permit of a detailed account of the return voyage, but on the 20th of March they reached the camp on the Murrumbidgee from which they had started. The relief party were not there, and there was nothing left but to toil on, though the men were falling asleep at the oars, and the river itself rose and raged madly against them. When they reached a point within ninety miles of the depot where Sturt expected the relief party to be, they landed, and two men—Hopkinson and Mulholland—went forward on foot for succour. They were now almost utterly without food, and had to wait six dragging days before men arrived with drays and stores to their aid.

One little item let me add; the boat being no longer serviceable, was burnt, Sturt giving as a reason that he was reluctant to leave her like a log on the water. What a priceless relic that boat would now have become!

Sturt received but scant appreciation on his return from this heroic journey. His eyesight was impaired and his health was failing; but instead of obtaining much-needed rest, he was sent to Norfolk Island, with a detachment of his regiment. There the moist climate still further prejudiced his health, though he was able to quell a mutiny of the convicts, and to save Norfolk Island from falling into their hands. Governor Darling too proposed that Sturt should be sent as British Resident to New Zealand, but filled with the love of continental exploration, he would not leave Australia, to the satisfaction of the fossils of the Colonial Office, who did not know of him, and promptly appointed Busby. Even Sir G. Murray, after whom the river had been named, had never heard of the river.

In 1832 or a little later, the temporary loss of the sight of one eye forced him to go to England on leave, when he also bade adieu to his regiment, which was ordered to India.

While in England, he published the first of his maps and books, but his eyesight totally failing him, he retired from the army, July, 1833. Sturt's eyesight, although never the same as before, was gradually restored to him, and on September the 21st, 1834, he was married at Dover to Charlotte Greene.

We must now take leave of this distinguished man, until he reappears in these pages as an explorer of Central Australia.*

*See Chapter XII.

Chapter VII.

SIR THOMAS MITCHELL.

(i.)—Introductory.

Mitchell, whose name both as explorer and Surveyor-General looms large in our history, was born at Craigend, Stirlingshire, in 1792. He was the son of John Mitchell of Grangemouth, and his mother was a daughter of Alexander Milne of Carron Works. When he was but sixteen, young Mitchell joined the army of the Peninsula as a volunteer. Three years later he received a commission in the 95th Regiment or Rifle Brigade. He was employed on the Quartermaster General's staff at military sketching; and he was present in the field at Ciudad Rodrigo, Badajoz, Salamanca, the Pyrenees, and St. Sebastian. After the close of the war he went to Spain and Portugal to survey the battlefields. He received promotion to a Lieutenancy in 1813. He served in the 2nd, 54th, and 97th Regiments of foot, and was promoted to be Captain in 1822, and Major in 1826. His appointment as Surveyor-General of New South Wales, as successor to John Oxley, took place in 1827, when he at once assumed office, and started energetically to lay out and construct roads, then the urgent need of the new colony.

Sir Thomas Mitchell.

His strong personality, and the energy and thoroughness he displayed in all his undertakings, combined with his many gifts as draughtsman, surveyor and organizer, proved to be of peculiar service to the colony at that period of its existence. There was a vast unknown country surrounding the settled parts, awaiting both discovery and development, and Mitchell's inclinations and talents being strongly directed towards geographical discovery, the office of Surveyor-General that he held for so long was the most appropriate and advantageous appointment that could have been given him in the interests of the colony.

At the same time, Major Mitchell had faults which have always detracted from the estimation in which he would otherwise be held for his undoubted capabilities. His domineering temper led him into acts of injustice, and often made it impossible for him to allow the judgments of others to influence his opinions. In his view, no other explorer but himself ever achieved anything worthy of commendation or propounded any credible theory regarding the interior of Australia. He always referred slightingly to Sturt, Cunningham, and Leichhardt, and his perversity on the subject of the junction of the Darling and the Murray drew even from the gentle Sturt a richly-deserved and unanswerable retort. On his second expedition, which was supposed to establish the identity of the Darling with the junction seen by Sturt, Mitchell excused himself from further exploration of the lower Darling as he expressed himself satisfied that Sturt's supposition was justified. But later, when on his expedition to what is now the State of Victoria, he again fell into a doubting mood, and he was not finally convinced until he had re-visited the junction. This constant doubting at last roused Sturt, who speaking in 1848 of Mitchell's work, said:—"In due time he came to the disputed junction, which he tells us he recognised from its resemblance to a drawing of it in my first work. As I have since been on the spot, I am sorry to say that it is not at all like the place, because it obliges me to reject the only praise Sir Thomas Mitchell ever gave me."

Sturt's original sketch of the junction had been lost, and Sturt, who was nearly blind at the time of publication, obtained the assistance of a friend, who drew it from his verbal description.

(ii.)—The Upper Darling.

Rumours of a mysterious river called the Kindur, which was said, on no better authority than a runaway convict's, to pursue a north-west course through Australia, now began to be noised about. This convict, whose name was Clarke, but who was generally known as "the Barber," said that he had taken to the bush in the neighbourhood of the Liverpool Plains, and had followed down a river which the natives called the Gnamoi. He crossed it and came next to the Kindur. This he followed down for four hundred miles before he came upon the junction of the two. The union of the two formed a broad navigable river, which he still followed, although he had lost his reckoning, and did not know whether he had travelled five hundred or five thousand miles. One thing, however, he was convinced of, and that was that he had never travelled south of west. He asserted that he had a good view of the sea, from the mouth of this most desirable river, and had seen a large island from which, so the natives reported, there came copper-coloured men in large canoes to take away scented wood. The Kindur ran through immense plains, and past a burning mountain. As no one had invited him to stay in this delectable country, he had returned.

The story, which bore every evidence of having been invented to save his back, received a certain amount of credence, and Sir Patrick Lindesay, then Acting-Governor, gave the Surveyor-General instructions to investigate the truth of it. It was in this way that Mitchell's first expedition originated.

On the 21st of November, 1831, Mitchell left Liverpool Plains and reached the Namoi on the 16th December. He crossed it and penetrated some distance into a range which he named the Nundawar Range. He then turned

back to the Namoi, and set up some canvas boats which
he had brought to assist him in following the river down.
The boats were of no use for the purpose, one of them
getting snagged immediately, and it was clear that it
would be easier to follow the river on land. As the range
was not easy of ascent, he worked his way round the
end of it and came on to the lower course of Cunningham's
Gwydir, which he followed down for eighty miles. At
this point he turned north and suddenly came to the
largest river he had yet seen. Mitchell, ever on the alert
to bestow native names on geographical features—a most
praiseworthy trait in his character, and through the
absence of which in most other explorers, Australian
nomenclature lacks distinction and often euphony—
enquired of the name from the natives, and found it to
be called the Karaula. Was this, or was this not the
nebulous Kindur? The answer could be supplied only
by tracing its course; but its general direction and the
discovery and recognition of its junction with the Gwydir
showed that the Karaula was but the upper flow of Sturt's
Darling. Much disappointed, for Mitchell was intent
upon the discovery of a new river system having a
northerly outflow, he prepared to make a bold push into
the interior. Before he started, Finch, his assistant-
surveyor, arrived hurriedly on the scene with a tale of
death. Finch had been bringing up supplies, and during
his temporary absence his camp had been attacked by
the natives, the cattle dispersed, the supplies carried off,
and two of the teamsters murdered. All ideas of further
penetration into the new country had to be abandoned.
Mitchell was compelled to hasten back, bury the bodies of
the victims, and after an ineffective quest for the
murderers, return to the settled districts.

The journey, however, had not been without good
results. Knowledge of the Darling had been considerably
extended, and it was now shown to be the stream receiving
the outflow of the rivers whose higher courses
Cunningham had discovered. The beginning of the great
river system of the Darling may be said to have been
thus placed among proven data. Mitchell himself after-

wards showed himself an untiring and zealous worker in solving the identity of the many ramifications of this system.

(iii.)—THE PASSAGE OF THE DARLING.

His next journey was undertaken to confirm the fact of the union of the Darling and the Murray. Sturt himself was fully convinced that he had seen the junction of the two rivers when on his long boat voyage; but he had not converted every one, and Mitchell, with a large party was despatched to settle the question and make a systematic survey. Early in March, 1833, the expedition left Parramatta to proceed by easy stages to the head of the Bogan River, which had been partly traversed the year before by surveyor Dixon. It was during this expedition that Richard Cunningham, brother of Allan, was murdered by the natives. He had not been long in Australia, and had been appointed botanist to the expedition. On the morning of April 17th, he lost sight of the party, whilst pursuing some scientific quest, and as the main body were then pushing hurriedly over a dry stage to the Bogan River, he was not immediately missed. Not having any bush experience, he lost himself, and was never seen again. A long and painful search followed, but owing to some mischance, Cunningham's tracks were lost on the third day, and it was not until the 23rd of the month that they were again found. Larmer, the assistant-surveyor, and three men were sent to follow them up until they found the lost man. Three days later they returned, having come across only the horse he had ridden, dead, with the saddle and bridle still on. Mitchell personally conducted the further search. Cunningham's tracks were again picked up, and his wandering and erratic footsteps traced to the Bogan, where some blacks stated that they had seen the white man's tracks in the bed of the river, and that he had gone west with the "Myalls," or wild blacks.*

*Lieut. Zouch, of the Mounted Police, subsequently found the site of his death, and recovered a few bones, a Manilla hat, and portions of a coat. The account afterwards given by the natives was to the effect that the white man came to them and they gave him food, and he camped with them : but that during the night he repeatedly got up, and this roused their fears and suspicions, so that they determined to destroy him. One struck him on the back of the head with a nulla-nulla, when the others rushed in and finished the deadly work.

As is often the case with men lost in the bush, the unfortunate botanist, by wandering on confusing and contradictory courses, had rendered the work of the search party more tedious and difficult, thus sealing his

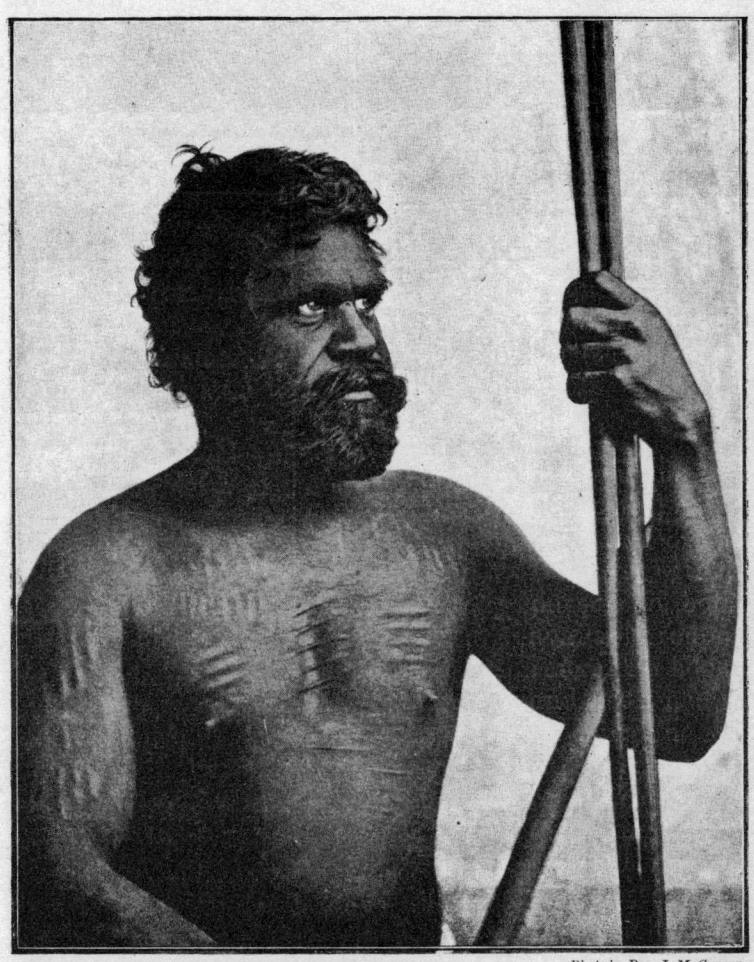

A Chief of the Bogan River Tribe.
Photo by Rev. J. M. Curran.

own fate. A rude stone memorial has since been erected on the spot, and a tablet put up in the St. Andrew's Scots Church, Sydney. The death of Cunningham, who was a young and ardent man with the promise of a brilliant future, caused Mitchell much distress of mind. He did all he could to find his lost comrade, and jeopardised the success of the expedition by the long delay of fourteen days.

He resumed his journey by easy stages down the Bogan, and on the 25th of May came to the Darling. This river was at once recognised by all who had been with him on his former trip as identical with the Karaula as Mitchell had supposed; but he found the country in a different condition from that presented by it when Sturt and Hume first discovered the river at nearly the same place. The water was now fresh and sweet to drink, and the flats and banks luxuriant with grass and herbage.

After choosing a site for a camp, where the town of Bourke now stands, Mitchell erected a stockade of logs, which he named Fort Bourke, after the Governor. The country on either side of the Darling was now alive with natives, and though a sort of armed truce was kept up, it was at the cost of constant care and watchfulness, and the tactful submission to numerous annoyances, including much petty pilfering. The boats proved to be of no service, and after Mitchell with a small party had made a short excursion down the river to the farthest limit of Sturt and Hume in 1829, where he saw the tree then marked by Hume, "H.H.," he had the camp dismantled, and started with the whole party to follow the river down to its junction with the Murray.

By the 11th of July, one month after leaving Fort Bourke, they had traced the river for three hundred miles through a country of level monotony unbroken by any tributary rivers or creeks of the least importance. Mitchell was now certain from the steadfast direction the river maintained, and the short distance that now intervened between the lowest point they had reached and Sturt's junction, that Sturt had really been correct in his surmise, and that he had witnessed the meeting of the rivers on that memorable occasion. He therefore decided that to keep on was but needlessly endangering the lives of his men. He was constantly kept in a state of anxiety for the safety of any member of the party whose duty compelled him to separate from the main body, for the natives, who had become doubly bold through familiarity, were now persistently encroaching and rapidly assuming a defiant manner.

On the very day that Mitchell had made up his mind to retreat, the long-threatened rupture took place. Mitchell refers to the blacks of this region as the most unfavourable specimens of aborigine that he had yet seen, barbarously and implacably hostile, and shamelessly dishonest. On the morning of July 11th, two of the men were engaged at the river, and five of the bullock-drivers were collecting their cattle. One of the natives, nick-named "King Peter" by the men, tried to snatch a kettle from the hand of the man who was carrying it, and on this action being resented, he struck the man with a nulla-nulla, stretching him senseless. His companion shot King Peter in the groin, and his majesty tumbled into the river and swam across. The swarm of natives who were constantly loitering around the camp, gathered together and advanced in an armed crowd, threatening the men, who fired two shots in self-defence, one of which accident-ally wounded a woman. Alarmed by the shots, three men from the camp came to the assistance of their mates, and one native was shot just when he was about to spear a man. The blacks now drew back a little, and the men seized the opportunity to warn the bullock-drivers, whom they found occupied in lifting a bullock that had fallen into a bog. Their arrival probably saved their lives, as the bullock-drivers were unarmed. No further attack took place, but the strictest watch had to be kept until the party was ready to begin the return journey or to beat a retreat as the natives regarded it. They reached Fort Bourke without further molestation, the aborigines being content with having driven away the whites, who retraced their steps from Fort Bourke to Bathurst.

The geographical knowledge gained on this journey consisted mainly in the confirmation of tentative theories —the identity of the Karaula with the Darling, and the uninterrupted course of the latter river southwards, as Major Mitchell himself had to confess, into the Murray.

Furthermore it seemed now satisfactorily settled that all the inland rivers as yet discovered found the same common embouchure. Mitchell's experience too proved that the pastoral country through which the Darling ran was by no means unfit for habitation, nor was the river a salt one; true some of his men had noticed that the water was brackish in places, but this brackishness, it was seen, had a purely local origin.

Mitchell was a keen observer of the habits and customs of the aborigines. He was remarkably quick at detecting tribal differences and distinctions, and his records of his intercourse with them—which occupies so much of his journals—were most interesting then, when little had been written on the subject; and are even more valuable now, as a first-hand account by an intelligent man and a practised observer of the appearance of the natives at the time of earliest contact with the white man.

(iv.)—Australia Felix.

One would have thought that the fact of the union of the Darling and the Murray was now sufficiently well-established; but the official mind deemed otherwise. When the Surveyor-General's next expedition started in March, 1836, he was informed that the survey of the Darling was to be completed without any delay; that, having returned to the point where his last journey had come to an end, he was to trace the river right into the Murray—see the waters of the two mingle in fact—then to cross over the Murray and follow up the southern bank, recrossing, and regaining the settled districts at Yass Plains. Although the primary object of the expedition was the verification of previous discoveries, the programme was largely departed from, and this particular journey of Mitchell's led to the opening up and speedy settlement of what is now the State of Victoria.

A drought, long-continued and severe, was in full force when Mitchell commenced his preparations for departure; consequently bullocks and horses in suitable condition were hard to obtain. But as the Government spared no expense, the necessary animals were at last available. Though upon reaching Bathurst Mitchell was informed that the Lachlan River was dry, he started on his third exploring expedition in the best of spirits. His mind overflowed with old memories and associations, and he wrote in his journal that this was the anniversary of the day "when he marched down the glacis of St. Elvas to the tune of 'St. Patrick's Day in the Morning,' as the sun rose over the beleaguered towers of Badajoz." He had heard that the aborigines of the lower Murray had been informed of his approach, and that they had assured the other tribes that they were gathering "murry coolah" —very angry—to meet him, but this to one of the Major's temper, lent but an added zest to the journey; for there were old scores to settle on both sides. It was the 17th of March, 1836, before he got free of the cattle stations and found himself at the point where Oxley had finally left the river. He noticed that throughout this route, in spite of the dry weather, the cattle were all in good condition; and he found Oxley's swamps and marshes transmuted into grassy flats. In fact, so changed was the face of the land, that even the landmarks of that explorer could scarcely be recognised.

Again his mind began to be troubled with doubts as to whether he had not acknowledged the veracity of Sturt's judgment too hastily, for we find in his journal that he again wavered, after professing that the identity admitted of little doubt. Now, on the Lachlan, he reverted to his old idea that the Darling drained a separate and independent basin of its own. He wrote:—

"I considered it necessary to ascertain, if possible, and before the heavy part of our equipage moved further forward, whether the Lachlan actually joined the Murrumbidgee near the point where Mr. Oxley saw its waters covering the face of the country, or whether it pursued a course so much more to the westward as to have been mistaken for the Darling by Captain Sturt."

Impelled by this doubt he undertook a long excursion to the westward with no result but the discomfort of several thirsty nights and an unchanging outlook across a level expanse of country bounded by an unbroken horizon. He reached Oxley's furthest on the 5th of May, but did not find that explorer's marked tree, though he found others marked by Oxley's party with the date 1817.

On the 12th of May, he halted on the bank of the Murrumbidgee, which in his opinion surpassed all the other Australian rivers he had yet seen. As his orders were simply to clear up the last hazy doubts that wrapped the Murray and Darling junction, and then to visit the southern bank of the Murray, he did not take his heavy baggage on to the Darling, but formed a stationary camp on the Murrumbidgee, and thence went on with a small party. When they came to the Murray, they found their old enemies awatch for them. It was afterwards ascertained that many of these aborigines had travelled as far as two hundred miles to assist in chasing back the white intruders once more from their violated hunting-grounds. But these braves of the Darling did not yet understand the nature of the man they sought to intimidate.

At first a nominal peace prevailed, and for two days the blacks followed the expedition closely, seeking to cut off any stragglers, and rendered the out-roving work of minding and collecting the cattle and horses one of considerable risk. Mitchell was soon convinced that a sharp lesson was necessary to save his men. In the event of losing any of his party, he would have had to fight his way back with the warriors of what seemed a thickly-populated district arrayed against him. One morning, therefore, the party was divided, and half of them sent back to an ambush in the scrub. The natives were allowed to pass on in close pursuit of the advance party. The native dogs, however, scented this ambuscade, and, after their fashion, warned the blacks of the presence of the hidden whites. As they halted, and began handling and poising their spears, one of the ambushed men fired without orders, and the others followed his example. The natives faltered, and those in advance, hearing the firing, rushed back eager to join in the fray. The conflict was

short and decisive: the over-confident fighting-men of the
Darling lost seven of their number and were driven
ignominiously back into the Murray scrub and across
that river. Henceforth the explorers were unmolested.
These pugnacious aboriginals were the same that
had threatened to bring Sturt's boat voyage to a tragical
conclusion, and soon after Mitchell's exploration, they
waged a determined war against the early overlanders
and their stock.

Mitchell's way to the Darling was now clear, and on the
31st of May he came upon that river, a short distance
above the confluence. Tracing the stream upwards, he
again convinced himself that it was the same river that
he had been on before, and, satisfied of this, he turned and
proceeded right down to the junction itself, and finally
disposed of one of the most interesting problems in
Australian exploration.

He naturally felt much anxiety, after his late skirmish,
for the safety of the stationary camp he had left behind,
and having lost no time during his return, he was relieved
to find his camp in quiet and safety.

The Surveyor-General first mapped the exact junction
of the Murrumbidgee and Murray, and then transferred
the whole of the expedition in boats to the other side
of the Murray. Thus was commenced the investigation
of the unexplored side of the Murray, that above its
junction with the Murrumbidgee, in other words the Hume
proper. On the 30th of June the party camped at Swan
Hill, having found the country traversed to exceed
expectations in every way. This pleasing state of affairs
continued and Mitchell journeyed on without check or
hindrance. After finding the Loddon River on the 8th
of July, and the Avoca on the 10th, he altered his
preconceived plan to follow the main river up, and, drawn
by the beauty and pastoral advantages of this new
territory, he struck off to the south-west in order to
examine it in detail, and trace its development southwards.

More and more convinced that he had found the garden
of Australia—he afterwards named this region

"Australia Felix"—Mitchell kept steadily on until he came to the Wimmera, that deceptive river which afterwards nearly lured Eyre to a death of thirst. On the last day of July he discovered the beautiful Glenelg, and launched his boat on its waters. At the outset he was stopped by a fall, was compelled to take to the land once more, and proceeded along the bank, occasionally crossing to examine the other side. On the 18th the boats were again used, the river being much broader, and in two days he reached the coast, a little to the east of Cape Northumberland.

The whole expedition then moved homewards, and reached Portland Bay, where they found that the Henty family from Van Diemen's Land had been established on a farm for about two years. From them Mitchell received some assistance in the way of necessary supplies, and then resumed his journey for home. On the 19th the party separated; Mitchell pushed ahead, leaving Stapylton, his second, to rest the tired animals for a while and then to follow slowly. On his homeward way Mitchell ascended Mount Macedon, and from the summit saw and identified Port Phillip. His return, with his glowing report of the splendid country he had discovered—country fitted for the immediate occupation of the grazier and the farmer—at once stimulated its settlement, and as the man whose explorations were of immediate benefit to the community in general—Mitchell's name stands first on the roll of explorers.

(v.)—Discovery of the Barcoo.

Some years elapsed before Mitchell—now Sir Thomas —again took to the field of active exploration. The settlement of the upper Darling and the Darling Downs had caused numerous speculations as to the nature of the unknown territory comprising the northern half of Australia. In 1841, communications had passed between the Governor and Captain Sturt, and in December of the same year Eyre, not long returned from his march round the Great Bight, wrote offering his services, provided that no prior claim had been advanced by Sturt. Governor Gipps asked for an estimate of the expenses, but considered Eyre's estimate of five thousand pounds too high, and nothing further was done. In 1843, Sir Thomas Mitchell submitted a plan of exploration to the Governor, who consulted the Legislative Council. The Council approved it and voted one thousand pounds towards expenses. The Governor referred the matter to Lord Stanley, whose reply was favourable, but the project still hung fire. In 1844 Eyre again wrote offering to make the journey at a much more reasonable rate, but his offer was however declined as Mitchell's proposals held the field. In 1845 the fund was increased to two thousand pounds, and Sir George Gipps ordered the Surveyor-General to make his preparations.

Mitchell favoured the search for a practicable road to the Gulf of Carpentaria, and hoped also that he would at last find his long-sought northern-flowing river. In a letter which he then received from a well-known grazier, Walter Bagot, there is mention of an aboriginal description of a large river running northward to the west of the Darling. But as natives in their descriptions frequently confuse flowing to and flowing from, they probably had Cooper's Creek in mind.

During the earlier part of the year, Commissioner Mitchell, the son of Sir Thomas, who was afterwards drowned during a passage to Newcastle, had made a flying survey towards the Darling, and the discovery of the Narran, Balonne, and Culgoa rivers has been attributed to him.

On the 15th of December, 1845, Mitchell started from Buree with a very large company, including E. B. Kennedy as second in command, and W. Stephenson as surgeon and collector. He struck the Darling much higher than Fort Bourke, and it was not until he was across the river that he passed the outermost cattle-stations, which had sprung rapidly into existence since his last visit to the neighbourhood. The Narran was then followed up until the Balonne was reached. This river, in his superlative style, Mitchell pronounced to be the finest in Australia, with the exception of the Murray. He then struck and followed the Culgoa upwards until it divided into two branches; he skirted the main one, which retained the name of the Balonne. On the 12th of April he came to the natural bridge of rocks which he called St. George's bridge, and which is the site of the present town of St. George. Here a temporary camp was formed; Kennedy was left in charge to bring the main body on more slowly; Mitchell with a few men went ahead. He followed up the Balonne to the Maranoa, but as the little he saw of that tributary did not tempt him to further investigation of it, he kept on his course up the main stream until he reached the junction of a stream which he named the Cogoon. This riverlet led him on into a magnificent pastoral district, in the midst of which stood a solitary hill that he named Mount Abundance. It is in his description of this region in his journal that we first find an allusion to the bottle tree.

The party wandered on over a low watershed and came down out on to a river which, from its direction and position, he surmised to be the Maranoa, the stream he had not followed. At this new point it was full of deep reaches of water, and drained a tract of most pleasing land. On its banks he determined to await Kennedy's arrival.

Kennedy overtook him on the 1st of June, bringing from Sir Thomas's son Roderick despatches which had reached the party after the leader's departure. Amongst other items of news in the despatches was the report of Leichhardt's return, and of the hearty reception that he had been accorded in Sydney. One piece of random information, a mere floating newspaper surmise, but enough to arouse Mitchell's suspicious temper, annoyed him greatly. "We understand," it ran, "the intrepid Dr. Leichhardt is about to start another expedition to the Gulf, keeping to the westward of the coast ranges."

As this seemed to indicate an intention of trespassing on Mitchell's present field of operations, he naturally felt some resentment not likely to be allayed by such a paragraph as the following:—"Australia Felix and the discoveries of Sir Thomas Mitchell now dwindle into comparative insignificance.''

Again leaving Kennedy, he set out to make a very extended excursion. Traversing the country from the head of the Maranoa, he discovered the Warrego River. Keeping north, over the watershed, for a time he fondly imagined that he had reached northward-flowing waters; but the direction of the rivers that he found, the Claude and the Nogoa, soon convinced him of his error, and that he was on rivers of the east coast. Even when he had reached the Belyando, a river which he named and followed down for a short distance, he still deluded himself that he had reached inland waters. Intensely mortified at finding that he was on a tributary of the Burdekin, and approaching the ground already trodden by Leichhardt, he returned to the head of the Nogoa, once more subdivided his party, and formed a stationary camp to await his return from a westward trip.

This time, however, he was blessed with the most splendid success. He found the Barcoo, a river that seemed to him to promise all he sought for. The direction of its upper course easily led him to believe that it was an affluent of the Gulf of Carpentaria, and after tracing it for some distance he returned to camp. The newly-discovered river he named the Victoria, thinking it would prove to be the same as that found by Captain Stokes on his survey expedition. It was on the Barcoo, or Victoria, that Mitchell first noticed the now famous grass that bears his name. On their return journey, they followed down the Maranoa, and at the old camp at St. George's Bridge, they were told by the natives that white men had visited the place during their long absence. It was a singular and welcome feature of Mitchell's discoveries that they had always proved to be adjacent to civilisation, and to be suitable for immediate occupation.

The discovery of the Barcoo was the last feather in the cap of the Surveyor-General. He was doomed to learn soon that it was not the river of his dreams, but only the head waters of that central stream discovered by Sturt, Cooper's Creek; but meanwhile the delusion must have been very gratifying.

In 1851 Mitchell was sent out to report on the Bathurst goldfields, and on a subsequent visit to England he took with him the first specimen of gold and the first diamond found in Australia. He was for a short time one of the members for the Port Phillip electorate, but resigned, as he found faithful discharge of the duties to be incompatible with his office. He patented the "boomerang screw propeller," and was the author of many educational and other works, including a translation of the Lusiad of Camoens. Although a strict martinet in his official duties, and subject to a choleric temper, he was strenuous in his devotion to the advancement of Australia, among whose makers he must always occupy a proud position. He died on the 5th of October, 1855, at "Carthona," his private residence at Darling Point, Sydney, New South Wales. His wife was a daughter of Colonel Blount.

Chapter VIII.

THE EARLY FORTIES.

(i.)—Angas McMillan and Gippsland.

Angas McMillan, who was the discoverer of what is now so widely-known as Gippsland, in Victoria, was a manager of the Currawang station, in the Maneroo district. On the 20th of May, 1839, he started from the station on a trip to the southward to look for new grazing land. He had with him but one black boy, named Jimmy Gibbu, who claimed to be the chief of the Maneroo tribe, so that if the party was small, it was very select. On the fifth day McMillan got through to the country watered by the Buchan River, and, from the summit of an elevation which he called Mount Haystack, he obtained a most satisfactory view over the surrounding region. The next night, McMillan, awakened by a noise, found Jimmy Gibbu bending over him with a nulla-nulla in his hand. Fortunately, McMillan's pistol was within easy reach, and, presenting it at Jimmy's head, he compelled him to drop the nulla-nulla, and to account for his suspicious attitude. Jimmy confessed to a fear of the Warrigals, or wild blacks of that region, to acute homesickness, and to a general unwillingness to proceed further.

McMillan examined the country he had found, and having judged it to be very desirable pastoral land, he returned home. He then formed a new station for Mr. Macallister on some country he had found on the Tambo River, and went himself on another trip of discovery. This time he had four companions with him, two friends named Cameron and Matthews, a stockman, and a black boy. They followed the Tambo River down its course through fine grazing country, both plains and forest, until in due course it led them to the point of its embouchure in the lakes of the south coast. He named Lake Victoria, and then directed his course to the west, where he discovered and named the Nicholson and Mitchell rivers. He was so deeply impressed with the resemblance of the country he had just been over to some parts of Scotland, that he called the district by the now obsolete name of "Caledonia Australis." On January the 23rd, 1840, he was out again and discovered and named the Macallister River, and pushed on as far west as the La Trobe River. This addition of rich pastoral regions to the already settled districts was altogether due to Angas McMillan's energy, and is now known as Gippsland, being named officially after Sir George Gipps, the Governor who had the amusing eccentricity of insisting that all the towns laid out during his term of office should have no public squares included within their boundaries, being convinced that public squares encouraged the spread of democracy.

(ii.)—COUNT STRZELECKI.

Count Strzelecki's expedition through Gippsland with the discovery of which district he is commonly and wrongly credited, was due to the literary and geographical work he had undertaken, as he was gathering material for his well-known work, "The Physical Description of New South Wales, Victoria, and Van Diemen's Land." He ascended the south-east portion of the main dividing range, and named the highest peak thereof Kosciusko, after a fancied resemblance in its outline to that Polish patriot's tomb at Cracow.

On the 27th of March, 1840, he reached the cattle station on the Tambo whither McMillan had just returned, and was directed by him on to his newly-discovered country. Strzelecki pushed through to Western Port, meeting with some scrubby and almost inaccessible country during the last stages of his journey. His party had to abandon both horses and packs, and fight its way through a dense undergrowth on a scanty ration of one biscuit and

a slice of bacon per day, varied with an occasional native
bear. It was here that the Count, who was an athletic
man, found that his hardy constitution stood the party in
good stead. So weakened and exhausted were his
companions, that it was only by constant encouragement
that he urged them along at all. When forcing their way
through the matted growth of scrub, he often threw himself bodily upon it, breaking a path for his weary
followers by the mere weight of his body. It was in a
wretched condition that they at last reached Western
Port.

(iii.)—Patrick Leslie.

In 1840 Patrick Leslie, who has always been considered
the father of settlement on the Darling Downs, started
with stock from a New England station, then the most
northerly settled district in New South Wales, and
formed the first station on the Condamine River, actually
before that river had been identified as a tributary of
the Darling. There was a general impression that the
Condamine flowed north and east, and finally found its
way through the main range to the Pacific. In 1841,
Stuart Russell, who closely followed Leslie as a pioneer,
followed the river down for more than a hundred miles
to the westward, and in the following year it was traced
still further, and the Darling generally accepted as its
final destination.

(iv.)—Ludwig Leichhardt.

Leichhardt is the Franklin of Australia, around whose
name has ever clung a tantalising veil of mystery
and romance. Truth to tell, his claim as a leading
explorer rests solely on his first and undoubtedly
fruitful expedition. But for his mysterious fate
mention of his name would not stir the hearts
of men as it does. Had he returned from his final venture
beaten, it must have been to live through the remainder
of his life a disappointed and embittered man. Far

better for one of his temperament to rest in the wilderness, his grave unknown, but his memory revered.

Leichhardt was born at Beskow, near Berlin, and studied at Berlin. Through an oversight he was omitted from the list of those liable to the one year of military service, and the sweets of exemption tempted him to evade the three-year military course. The consequence was that he was prosecuted as a deserter, and sentenced *in contumaciam*. Afterwards, Alexander von Humboldt succeeded, by describing his services to science on his first expedition in Australia, in obtaining a pardon from the King. By a Cabinet Order, Leichhardt received permission to return to Prussia unpunished. When the order arrived in Australia, he had already started on his last expedition.

Ludwig Leichhardt.

Dr. Leichhardt appears to have been a man whose character, to judge from his short career, was largely composed of contradictions and inconsistencies. Eager for personal distinction, with high and noble aims, he yet lacked that ready sympathy and feeling of comradeship that attract men. Leichhardt's followers never desired to accompany him on a second expedition. Yet strange to say, he was capable of inspiring firm friendship in such men as William Nicholson and Lieutenant Robert Lynd.

When he left on his first exploring expedition, on which he was successful owing to the luck of the novice, people generally predicted—and with much reason—that he would fail. But when he set out on his second and disastrous journey, universally applauded and with his name on everybody's lips, it was never doubted but that he would succeed.

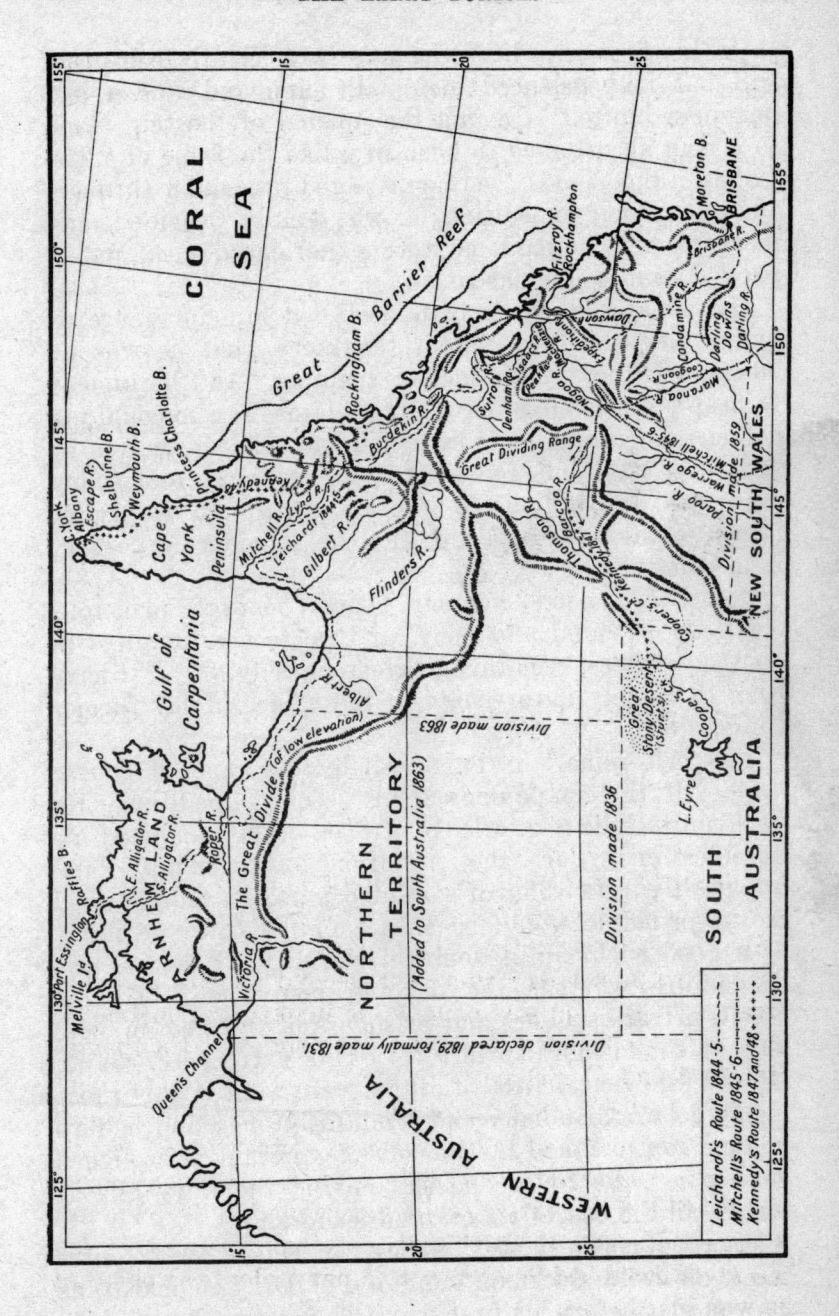
CORAL SEA
Great Barrier Reef
Rockingham B.
Princess Charlotte B.
Weymouth B.
Shelburne B.
Escape R.
Albany
C. York
Cape York Peninsula
Kennedy 1848
Mitchell R.
Lynd R.
Leichardt 1844-5
Gilbert R.
Flinders R.
Burdekin R.
Suttor R.
Denham Ra.
Peak Ra.
Isaacs R.
Mackenzie R.
Nogoa R.
Expedition Ra.
Dawson R.
Fitzroy R.
Rockhampton
Moreton B.
BRISBANE
Brisbane R.
Condamine R.
Darling Downs
Darling R.
Cogoon R.
Maranoa R.
Mitchell 1845-6
Warrego R.
Paroo R.
Great Dividing Range
Thomson R.
Barcoo R.
Kennedy 1847
Cooper's Ck.
Great Stony Desert (Sturt's)
L. Eyre
Division made 1859
NEW SOUTH WALES
SOUTH AUSTRALIA
Gulf of Carpentaria
Albert R.
The Great Divide (of low elevation)
Division made 1863
NORTHERN TERRITORY
(Added to South Australia 1863)
Division made 1836
Roper R.
ARNHEM LAND
E. Alligator R.
S. Alligator R.
Raffles B.
Port Essington
Melville Id
Victoria R.
Queen's Channel
Division declared 1829, formally made 1831.
WESTERN AUSTRALIA
Leichardt's Route 1844-5
Mitchell's Route 1845-6
Kennedy's Route 1847 and 48

On his first expedition he was insufficiently equipped, had but inexperienced men with him, and was a bad bushman himself. In fact the journal of the trip reads to a man accustomed to bush life, like the fable of "The Babes in the Wood;" yet he managed to blunder through. On his second expedition he was amply provided, and most of his companions were experienced men, but it proved a miserable fiasco.

His great confidence in himself led him to ignore or undervalue the fact, patent to others, that he was no bushman either by instinct or training. And he seemed to prefer for companions men like himself, who could not detect this failing, as is evident from a letter written by him to W. Hull, of Melbourne, with reference to a young man who was anxious to join his party. In this letter he enumerates the qualities that he considers necessary in a follower:—

"Activity, good humour, sound moral principle, elasticity of mind and body, and perfect willingness to obey my orders, even though given harshly. . . I have been extremely unfortunate in the choice of my former companions."

The last remark is an unworthy one, and of course applies to the companions of his second expedition. He does not include a knowledge of open-air life amongst his qualifications, nor the needful bushmanship; and apparently in Leichhardt's opinion, a useless man of good moral principle would be as acceptable to an explorer as a good bushman of doubtful morality. It causes one to inquire whether the devoted men who toiled for Sturt, private soldiers and prisoners of the Crown, were men of sound moral principle? This extract affords an insight into Leichhardt's failures. He wanted only those men who would blindly and ignorantly obey and believe in him. For a man of Leichhardt's temperament, such men were not to be found: he had missed the fairy gift at birth—all the essentials of good leadership.

Stuart Russell, in his "Genesis of Queensland," cites his shrewd old stockman's opinion of Dr. Leichhardt, as he was just before his first trip. The station from which

Leichhardt started on that occasion was near Russell's, so that the man spoke from personal knowledge:—"It's my belief that if Dr. Leichhardt do it at all, 'twill be more by good luck than management. Why, sir, he hasn't got the knack of some of us; why it comes like mother's milk to some. I can't tell how or why, but it does. Mark my words, sir, Dr. Leichhardt hasn't got it in him, and never will have."

Two invaluable qualities in an explorer, apart from his scientific attainments, Leichhardt possessed. These were courage and determination; necessary no doubt, but not sufficient in themselves to carry through an expedition to success. He lacked tact, and was deficient in practical knowledge of the bush, and especially in what is known as bushmanship. One fixed idea of his was, that in dry country if one can only keep on far enough one is bound to come to water: a theory plausible enough if it could be carried out to its logical conclusion; but the application of which often involves a physical impossibility. And it must be taken into consideration that Leichhardt had never travelled in the dry country of the interior, but that what small experience he possessed had been gained on the fairly well-watered coast. He asserts in his journal that cattle and horses trust entirely to the sense of vision for finding water, and not to the sense of smell. The exact reverse is of course the case.

The character of the lost explorer will thus be seen to have militated strongly against his success when he came to be pitted against the—to him—unknown dangers of a dry season in the far interior. But his fatal self-confidence led him to challenge the desert, thinking that he must succeed where better men had been denied even the hope of success. When his last expedition comes to be reviewed, a more detailed discussion of the probabilities of a successful issue to it will be made. Poor Leichhardt, with all his moods and caprices, it would have been strange if he had not shown some appreciation of humour. Let us quote his description of his sudden and unexpected arrival in Sydney, after the Port Essington expedition.

"We did come to Sydney, it was quite dark; we did go ashore, and then I thought to see my dear friend Lynd. So I went up George Street to the barracks. And then I went to his quarters to his window. He was dressing himself; I did put in my head; he did jump out of the other window and I stood there wondering. Soon many people did come round, and did look, Oh so timid. I did not know all. And there was such a greeting. I was dead, and was alive again. I was lost, and was found."

But in thus reviewing Leichhardt's aptitude—or rather inaptitude—for the work, and commenting upon his shortcomings, we must do him the fullest justice by paying homage to the sincerity of his belief in himself and his mission. In that belief he was honestly loyal. His conception of his duty was of the highest, and in its interest he would, and did, make every sacrifice in his power. If some prescient tongue could have told Leichhardt that the end of his quest would be an unknown death, he would have accepted the fate without a murmur, provided his death benefited geographical discovery.

As the man of science in a party under a capable leader, Leichhardt would have achieved greater success than many men who have filled that position; as the leader himself he was, of necessity, an absolute failure.

Leichhardt arrived in New South Wales in 1842, and, after some botanical excursions about the Hunter River district, he travelled overland to Moreton Bay, and there occupied himself with short expeditions in the neighbourhood, pursuing his favourite study of physical science. When the subject of the exploration of the north was mooted, he was desirous of securing the position of naturalist, but the delay in forming the projected expedition disappointed him, and he resolved to try and organise a private one. In this he received very little encouragement. He persevered, however, and eking out his own resources by means of private contributions, both in money and stock, he managed to get a party together. On the 1st of October, 1844, he left Jimbour station on the Darling Downs, on the trip that was destined to make his name as an explorer. His preparations were on a much smaller scale than Mitchell's. Considering the importance of the undertaking, his party was absurdly small. He had with him six white and two black men, seventeen horses, sixteen head of cattle and four kangaroo dogs; and his supply of provisions was equally meagre. His plan of starting from Moreton Bay to Port Essington differed considerably from Mitchell's proposed journey to the Gulf from Fort Bourke, but although longer and more roundabout, it would be a safer route for his little party to adopt, as they would keep to the comparatively well-watered coastal lands. Leaving the Condamine, he crossed the northern watershed, and struck the head of one of the main tributaries of the Fitzroy River, which he named the Dawson. Thence he passed westward into a region of fine pastoral country, which he named the Peak Downs. Here he named the minor waters of the Planet and the Comet, and Zamia Creek. On the 10th of January, 1845, he found the Mackenzie River, and thence crossed on to and named the Isaacs, a tributary of the Fitzroy coming from the north. This river they followed up till they crossed the watershed on to the head waters of the Suttor River. They followed this stream down until it brought them to the Burdekin, Leichhardt's most important discovery.

Up the valley of this river they travelled, until they reached the head, where, at the Valley of Lagoons, they crossed the watershed on to the waters of the Gulf of Carpentaria. Here, for some unknown reason, Leichhardt went far too much to the north, which necessitated a long detour around the south-eastern corner of the Gulf. It was while they were retracing a southern course along the eastern shore of the Gulf that the naturalist Gilbert met his fate. Up to this time they had been so little troubled with the natives that they had ceased almost to think of a possible hostile encounter with them. This fancied immunity was broken in a most tragic manner on the night of the 28th of June, 1845. It was a calm, quiet evening, and the party were peacefully encamped beside a chain of shallow lagoons. The doctor was thinking out his plans for the next few days, Gilbert was planting a few lilies he had gathered, as was his nightly habit when any flowers were available, Roper and the others were grouped around the fire warding off the attacks of the mosquitoes. Suddenly about seven o'clock a shower of spears was thrown among the unarmed men, and Gilbert was almost instantly killed, Roper and Calvert being seriously wounded. The whites rushed for their guns, but unfortunately not one weapon was ready capped, and it was some time before any of them could be discharged, when a volley caused the blacks to scamper off. It is most astonishing that the whole of the members of the party were not cut down in one dreadful massacre.

The body of the murdered naturalist was buried at the fatal camp, but the grave was left unmarked, and a large fire built and consumed above it to hide all traces of it from the natives. The river where this sad mishap occurred now bears the name of Gilbert.

From the scene of this tragedy, which ordinary precautions would have avoided, the party proceeded around the southern shore of the Gulf, keeping a short distance above tidal waters; but their progress was slow and painful on account of the two wounded men. Most of Leichhardt's names are still retained for the rivers of the Gulf which he crossed, the Leichhardt itself being an exception. This river he mistook for the Albert, so named by Captain Stokes during his marine survey of the north coast. A. C. Gregory rectified the error in after years, and gave the river the name of the lost explorer for whom he was then searching. With fast-dwindling supplies, lagging footsteps, and depressed spirits, the expedition travelled slowly on to the south-west corner of the Gulf where, in crossing a large river, the Roper, four of the horses were drowned in consequence of the boggy banks. This misfortune so limited their means of carriage that Leichhardt had to sacrifice the whole of his botanical collection. On the 17th of December, 1845, the worn-out travellers, nearly destitute of everything, reached the settlement of Victoria, at Port

Essington, and the long journey of fourteen months was over.

This expedition, successful as it was in opening up such a large area of well-watered country, attracted universal attention both to the gratifying economic results and to the hitherto untried leader. He was enthusiastically welcomed back to Sydney, and dubbed by journalists the prince of explorers. But what captivated public fancy was a certain halo of romance that clung to the journey on account of the reported death of Leichhardt, a report that gained general credence. His unexpected return invested him with a romance which—fortunately for his reputation—the total and absolute disappearance of himself and company in 1848 has but the more richly coloured. Enthusiastic poets gushed forth in song, and a more substantial reward was raised by public and private subscriptions and shared among the expedition in due proportions.

Encouraged by these encomiums on his success, and perhaps a little intoxicated by the general acclamation, Leichhardt now conceived the ambitious idea of traversing the continent from the eastern to the western shore; keeping as far as possible on the same parallel of latitude. This was a bold project, coming as it did so soon after Sturt had returned to Adelaide from his excursion into the interior with a terrible tale of thirst and suffering. But this time the hero of the hour experienced no difficulty in obtaining funds and other necessary aids. The party, when organised, travelled from the Hunter River to the Condamine, taking with them their outfit of mules, cattle, and goats. When the expedition departed from Darling Downs, they numbered seven white men and two natives, with 270 goats, 180 sheep, 40 bullocks, 15 horses, and 13 mules. There were besides an ample outfit and provisions calculated to last the explorers on a two years' journey; for it was estimated that the expedition would be absent from civilisation for that time.

Instead of setting out westwards from the initial point in a direction where Leichhardt could reasonably expect fair travelling country for some distance, he proceeded along his old track north to the Mackenzie and Isaacs Rivers. What induced him to adopt this course is uncertain. He explained to one of his party that it was to verify some former observations; or he may have had some dim notion that by keeping to the tropical line he would gain some climatic assistance. Whatever the cause, the result was disastrous. The wet season and monsoonal rains caught the party amongst the sickly acacia scrubs of that region; and hemmed in by mud and bog they lost their stock, consumed their provisions, and made no progress. Henceforth the narrative is one of semi-starvation, varied by gorging on the days when a beast was killed; and wrangles and quarrels, in which the leader appeared in no amiable light. Medicine had been omitted from the stores, and all the covering they had from the torrential rains was provided by two miserable calico tents. The 6th day of July found them back on Chauvel's station on the Condamine; a sad contrast to the party which had aspired to cross the continent.

The onus of this wretched failure Leichhardt tried to cast upon his companions, upon whom he made many unjust aspersions. J. F. Mann, late of the Survey Department of New South Wales, was one of the expedition, and the last surviving member of any expedition connected with Leichhardt. He wrote a booklet in which he vigorously defends his comrades and himself against the unworthy slurs cast at them by Leichhardt. Amongst his papers is a rough sketch from life of Leichhardt in bush costume.

On reaching the Condamine, Leichhardt was put into possession of the news of Mitchell's return and of the discovery of the Barcoo. Being anxious to examine the country lying between the upper Condamine and Mitchell's latest track, he, in company with two or three of his late companions, left Cecil Plains for that purpose; he went as far as the Balonne River, crossed it and returned. This doubtless was in view of organising another expedition, with which he evidently intended to start in another manner, straight to the westward.

Still persisting and believing in his capability of leading an expedition across the continent, and fearful that this ambitious project might be forestalled, he now made strong and strenuous efforts to organise another party. He succeeded at length, but the party was neither so well provided, nor so large, nor composed of such capable men as the second.

In fact, very little is known of the members that composed it; the only thing certain is that it was not at all adapted for the work that lay before it. A few words of the Rev. W. B. Clarke, the well-known geologist, have been many times quoted, and they convey about all that is known of the personnel of the expedition:—

John Frederick Mann. Born 1819, died September 7th, 1907, at Sydney. The last survivor of a Leichhardt expedition.

"The parties that accompanied Leichhardt were perhaps little capable of shifting for themselves in case of any accident to their leader. The second in command, a brother-in-law of Leichhardt, came from Germany to join him before starting, and he told me, when I asked him what his qualifications for the journey were, that he had been at sea and had suffered shipwrecks, and was therefore well able to endure hardship. I do not know what his other qualifications were."

The last sentence is very pregnant, and implies that a very poor opinion of the men as experienced bushmen was entertained by those who saw them.

The lost expedition is supposed to have consisted of six whites and two blacks; the names known being those of the doctor himself, Classen, Hentig, Stuart, and Kelly. He had with him 12 horses, 13 mules, 50 bullocks, and 270 goats; beside the utterly inadequate allowance of 800 pounds of flour, 120 pounds of tea, some sugar and salt, 250 pounds of shot, and 40 pounds of powder. His last letter is dated the 3rd of April, 1848, from McPherson's station on the Cogoon, but in it he speaks only of the country he has passed through, and nothing of his intended route. Since the residents of this then outlying station lost sight of him, no sure clue as to the fate of him and his companions has ever come to light. The total evanishment, not alone of the men, but of the animals—especially the mules and the goats—is one of the strangest mysteries of our mysterious interior. Thirst probably caused the death of the animals, and in that case they would have died singly and apart, and their remains would in after years elude attention. A similar fate probably befel the men.

Rumour has always been rife as to the locality of Leichhardt's death, and suggestions the most hopelessly unlikely and inconsistent have been put forward and seriously considered. At the same time, the only two reliable marks, undoubtedly genuine and fitting in in every way with Leichhardt's projected course of travel, have been neglected.

Leichhardt started from McPherson's station on the Cogoon, now perhaps better known as Muckadilla Creek. There was a rumour, never authenticated, that after he had proceeded nearly one hundred miles he sent back a man with a report that he had passed through some splendid pastoral land, but this is not at all likely to be true. The first indication of him is then met with on the Barcoo (Victoria) whereon A. C. Gregory, in charge of the Leichhardt Search Expedition, in 1858, found his marked tree and other indications:—

"Continuing our route along the river (latitude 24 degrees 35 minutes; longitude 36 degrees 6 minutes), we discovered a Moreton Bay ash, about two feet in diameter, marked with the letter L on the east side, cut through the bark about four feet from the ground, and near it the stumps of some small trees that had been cut with a sharp axe, also a deep notch cut in the side of a sloping tree, apparently to support the ridge-pole of a tent, or some similar purpose; all indicating that a camp had been established here by Leichhardt's party. No traces of stock could be found; this however is easily accounted for, as the country had been inundated last season."

There can be little doubt about the authenticity of the trace, and it at once does away with the truth of the stories told to Hovenden Hely by the blacks as to Leichhardt's murder on the Warrego River. Gregory then went up the Thomson River but found no other mark, and returning, followed that river and Cooper's Creek down to South Australia. This camp of Leichhardt's is easily understood. Then follows an account of the other found by the same explorer in 1856, during an earlier expedition. This was on the upper waters of Elsey Creek, and his description of it runs as follows:—

"The smoke of bush fires was visible to the south, east, and north, and several trees cut with iron axes were noticed near the camp. There were also the remains of a hut, and the ashes of a large fire, indicating that there had been a party encamped there for several weeks; several trees from six to eight inches in diameter had been cut down with iron axes in fair condition, and the hut built by cutting notches in standing trees and resting a large pole therein for a ridge. This hut had been burnt apparently by the subsequent bush fires; and only some pieces of the thickest timber remained unconsumed. Search was made for marked trees, but none were found, nor were there any fragments of iron, leather, or other material of the equipment of an exploring party, or of any bones of animals other than those common to Australia. Had an exploring party been destroyed there, there would most likely be some indications, and it may therefore be inferred that the party proceeded on its journey. It could not have been a camp of Leichhardt's in 1845, as it is 100 miles south-west of his route to Port Essington, and it was only six or seven years old, judging by the growth of the trees; having subsequently seen some of Leichhardt's camps on the Burdekin, Mackenzie, and Barcoo Rivers, a great similarity was observed in the mode of building the hut, and its relative position with regard to the fire and water supply, and the position with regard to the great features of the country was exactly where a party going westward would first receive a check from the waterless tableland between the Roper and Victoria Rivers, and would probably camp and reconnoitre before attempting to cross to the north-west coast."

Leichhardt's track, as far as the Elsey, seems tolerably plain and entirely in accordance with the character of the man and his intentions. Forced to retreat from the dry country west of the Thomson, he probably followed that river to its head, and crossing the main watershed regained and re-pursued his track of 1845, as far as the Roper, of which river Elsey Creek is a tributary. When he left the camp seen by Gregory, he would, going either south-west or west, find himself in the driest of dry country, which is even now but sparsely settled. And there came the end.

Long before the last water they carried with them had been used, their beasts would have all died, left here and there wherever they fell. So too would the men. Differences of opinion would have arisen, and some would have been for turning back, and others for keeping on. Some would have persisted in changing the direction they were following, and, led on by some mad delirious fancy in seeing water indications in some rock or bush, would have separated and staggered on to die alone.

Their baggage would have been left strewn over the desert where it had been abandoned, and the men, one by one, would have shared the same fate. Into such a waterless and barren region the blacks would seldom penetrate, and what with the sun, hot winds, bush fires, and sand-storms, all recognisable traces would soon have been effaced.

With regard to the notched tree to support a ridge-pole, which feature was noticed by Gregory in both camps, J. F. Mann, of whose companionship with Leichhardt mention has already been made, often stated that he would recognise Leichhardt's camps anywhere, by this singular device for supporting the ridge of a tent.

Chapter IX.

EDMUND B. KENNEDY.

(i.)—The Victoria and Cooper's Creek.

E. B. Kennedy, whose tragic death ineffaceably branded the Cape York blacks as remorselessly cruel, came to Australia early in life, and was appointed a Government surveyor in 1840. His first experience as an explorer was gained when as Assistant-Surveyor and second in command he accompanied his chief on the last expedition that Mitchell led into the interior. On this occasion he remained in charge of the camp formed at St. George's Bridge, and then conducted part of the expedition on to the Maranoa, where he rejoined the Major, and remained in charge whilst Mitchell made his exploration westward.

Edmund B. Kennedy.

On Mitchell's return to Sydney, there being some doubt as to the point of outflow of the newly-discovered Victoria River, Kennedy was sent out with a small party to follow the river down and ascertain its course and destination.

On the 13th of August, he reached Mitchell's lowest camp on the Victoria River, and started to trace the river down. During the first day's journey he came across some natives, from one of whom he learnt that the aboriginal name of the river was the Barcoo. Two days afterwards he observed with some anxiety that the trend of the valley was inclining from northwards towards the point whence Sturt had turned back from his upward course on Cooper's Creek. As the second part of his instructions was to find a practicable road to the Gulf, he feared that he would not have sufficient provisions to fulfil both duties. He therefore made a stationary camp, and with two men proceeded down the river. But after two days' journey, he found that the Barcoo turned to the west, and even north of west. The channel now showed large reaches of water within its confines, some of them more than one hundred yards in width. This induced him to alter his plan, and he thought he should follow such an important watercourse and ascertain its outflow. He therefore turned back for the remainder of his party. On the 30th of August he discovered a large river coming from the N.N.E., and he named it the Thomson. With the usual inconsistency of Australian inland rivers, the Thomson soon presented another and different scene. The great pastoral stretches of the upper course were left behind, and were succeeded by flat and inferior country intersected by sand-ridges. The course of the river itself once more turned to the southward, and was but scantily watered. Still Kennedy persevered until convinced that further progress must bring him to Sturt's furthest on Cooper's Creek. The face of the land answered to Sturt's description; and grass and feed both beginning to fail him, Kennedy had to consider whether it was worth while risking the lives of his men to confirm what was practically a certainty. At last vistas of the desert, described by Sturt with such terrible fidelity, appeared stretching away to the horizon, and Kennedy turned back, satisfied that the Victoria River and Cooper's Creek were one and the same stream.

It was now Kennedy's intention to make an excursion towards the Gulf of Carpentaria. On his way down, in order to travel lighter, he had buried a large quantity of flour and sugar as well as his drays. When he arrived at the cache of provisions on his way back, he found that the natives had dug the rations up, and in mere wantonness had so mixed and scattered them as to render them useless. A little further on, he was just in time to save the carts, for an aboriginal was probing in the ground with a spear to ascertain their whereabouts. During this excursion Kennedy noticed that the blacks were given to "chewing tobacco in a green state;" but the "tobacco" was, of course, the pituri plant, which they are accustomed to masticate. By the time he reached the head of the Warrego, Kennedy was too short of provisions to attempt his projected Gulf expedition, and had to make homeward, but resolved to go down by that river and ascertain whether it joined the Darling or flowed westward.

The Warrego dividing into many dry channels when they reached its lower courses, the party struck eastward to the Culgoa, and reached that river after a very distressing stage over dry country on which they lost six horses from heat and thirst, whilst bringing the carts across it.

(ii.)—A Tragic Expedition.

Kennedy's first experience of an independent exploring expedition in the west was by no means a fitting prelude to the tragic journey he next undertook. The same impulse that led to Mitchell's and Leichhardt's northern journeys stimulated Kennedy to make his dangerous journey up the eastern coast of the long peninsula that terminates in Cape York—the desire to find a road to the north coast, so that an easy chain of communication should exist between the southern settlements and the far north.

It was at the end of the month of May that Kennedy landed at Rockingham Bay with his party of twelve men. He had started from Sydney in the barque "Tam o' Shanter," which was convoyed by Captain Owen Stanley in the "Alligator." This was in 1848, the same fateful year that witnessed Leichhardt's disappearance. A schooner was to meet the party on the north, at Port Albany, where it was proposed to form a settlement should the features of the peninsula warrant such an enterprise. In actual point of distance the task was not great, being a land traverse of from three to four hundred miles, allowing for deviations. But never were men in Australia so dogged by disaster and beset by danger as were Kennedy and his followers. Opposed by country as yet unfamiliar to them, they found their onward path hindered by many totally unforeseen conditions. Ranges and ravines clothed with an almost impenetrable jungle, which was infested with the venomous leaves of the stinging tree and the hooked spikes of the lawyer vine, confronted them. The land was densely populated with the most savage and relentless natives on the continent, who resented the invasion from the outset. Death tracked them steadily throughout, and claimed ten out of the thirteen of the devoted party as his victims.

The country through which their course lay is now dotted with mining-fields and townships, and fertile spaces of tilled tropical plantations. The coast-line rich in harbours is the busy haunt of steamers, and the narrow waterway between the mainland and the great barrier reef the home of many lightships. But when Kennedy and his party made their pioneer journey, the great desolation of the wilderness beset them on every side from the land, whilst the sea off-shore held myriad dangers.

Kennedy landed from the "Tam o' Shanter" at the little point that still bears the jovial name, and bade farewell to Owen Stanley in good spirits, and with no dread premonitions. He was fresh from the sun-scorched plains of the interior, and would confidently confront whatever might lie before him. Scrub and swampy country delayed him on his way to the higher land at the foot of the range, where he had hoped to find better travelling country; but the foothills were serried with ravines and gullies, and the sides clothed with the ever-present jungle. The horses and sheep, unaccustomed to the sour grasses of the coast lands of northern Australia, pined and rapidly wasted away. Their troubles were augmented by acts of annoyance, and on one unfortunate occasion, of open hostility on the part of the blacks.

By the 18th of July, a little over six weeks after they had left Rockingham Bay, the sheep had been reduced from one hundred to fifty, and the horses began to fail so rapidly that they had to abandon the carts, while the men were becoming completely exhausted from the endless cutting and hacking of the scrub. At length they surmounted the range, the backbone of the peninsula, and on the western slope, amid the heads of the rivers flowing into the Gulf of Carpentaria, made better progress. Kennedy, however, adhered to his instructions to examine the eastern slope, and recrossed the watershed, where troubles again came thick upon him. One after another the horses began to give in, and owing to the storekeeper's mismanagement, they were nearly out of provisions. On the 9th of December they reached Weymouth Bay, and Kennedy determined to form a stationary camp, and leaving there the main body of his men, push forward to Port Albany, whence he would send back the schooner that was awaiting them with relief. He selected seven men whom he left in charge of Carron, the naturalist, and with three men and the heroic Jacky-Jacky, an aboriginal of New South Wales, he pushed on—to his death.

Before the departure the last sheep was slaughtered, and its lean and miserable carcase shared between the two parties; and with Carron, Kennedy ascended a hill that commanded a prospect of the country lying to the north, but could see nothing but rugged hills and black scrub. He confided only to Carron his gloomy foreboding that he would never reach Albany, so disheartened were both the men by the prospect. And throughout those long weeks of starvation that ensued, Carron refrained from crushing all hope in his comrades by communicating to them Kennedy's despair of relief.

For three weeks Kennedy struggled on, cutting his path through the scrub, and, with dwindling strength, clambering across the spurs of the range. For the story of his struggles and eventual death Australia has had to rely on the report of the only survivor, the faithful Jacky-Jacky. They reached Shelburne Bay, where one of the men accidentally shot himself, and became so weak from loss of blood that it was impossible for him to move. As another man, Luff, was sick, Kennedy left the third man, Dunn, to attend to his two comrades, and pushed on alone with the native boy. He had actually gained the Escape River, within sight of Albany Island, when his fate overtook him, and, surrounded by the blood-thirsty foes who had so long and persistently hung upon his footsteps, he fell at last beneath their spears.

The story is best told in Jacky's own words, although it has been often repeated. They had come across some natives whom Kennedy was inclined to trust, but of whom Jacky was suspicious, and that night they camped in the scrub, foodless and fireless.

"I and Mr. Kennedy," said Jacky, "watched them that night, taking it in turns every hour that night. By and by I saw the blackfellows. It was a moonlight night, and I walked up to Mr. Kennedy and said: 'There is plenty of blackfellows now;' this was in the middle of the night. Mr. Kennedy told me to get my gun ready.

"The blacks did not know where we slept, as we did not make a fire. We both sat up all night. After this daylight came and I fetched the horses and saddled them. Then we went a good way up the river, and then we sat down a little while, and then we saw three blackfellows coming along our track, and then they saw us, and one ran back, as hard as he could run, and fetched up plenty more, like a flock of sheep almost. I told Mr. Kennedy to put the saddles on the horses and go on, and the blacks came up and they followed us all day. All along it was raining. I now told him to leave the horses and come on without them, that horses made too much track. Mr. Kennedy was too weak, and would not leave the horses. We went on this day until the evening; raining hard and the blacks followed us all day, some behind, some planted before. In fact, blackfellows all round following us. Now we went into a little bit of scrub, and I told Mr. Kennedy to look behind always. Sometimes he would do so, and sometimes he would not do so to look out for the blacks. Then a good many blackfellows came behind in the scrub and threw plenty of spears, and hit Mr. Kennedy in the back first. Mr. Kennedy said to me: 'Oh Jacky! Jacky! shoot 'em! shoot 'em!' then I pulled out my gun and fired and hit one fellow all over the face with buck-shot. He tumbled down and got up again and again, and wheeled right round, and two blacks picked him up and carried him away. They went a little way and came back again, throwing spears all round, more than they did before—very large spears.

"I pulled out the spear at once from Mr. Kennedy's back, and cut the jag with Mr. Kennedy's knife. Then Mr. Kennedy got his gun and snapped, but the gun would not go off. The blacks sneaked all around by the trees, and speared Mr. Kennedy again, in the right leg above the knee a little, and I got speared in the eye, and the blacks were now throwing always, never giving over, and shortly again speared Mr. Kennedy again in the right side. There were large jags in the spears, and I cut them off and put them in my pocket. At the same time we got speared the horses got speared too, and jumped and bucked about and got into the swamps. I now told Mr. Kennedy to sit down while I looked after the saddle-bags, which I did, and when I came back again I saw the blacks along with Mr. Kennedy. I then asked him if he saw the blacks with him. He was stupid with the spear wounds, and said 'No'; I then asked him where was his watch? I saw the blacks taking away watch and hat as I was returning to Mr. Kennedy. Then I carried Mr. Kennedy into the scrub. He said, 'Don't carry me a good way.' Then Mr. Kennedy looked this way, very bad (Jacky rolling his eyes). I asked him often, 'are you well now?' and he said—'I don't care for the spear wound in my leg, Jacky, but for the other two spear wounds in my side and back, and I am bad inside, Jacky!' I told him blackfellow always die when he got spear wound in there (the back). He said: 'I am out of wind, Jacky.' I asked

Wild Blacks of Cape York signalling

him: 'Are you going to leave me?' And he said, 'Yes, my boy; I am going to leave you; I am very bad, Jacky, you take the books, Jacky, to the Captain, but not the big ones; the Governor will give you anything for them.' I then tied up the papers. He then said: 'Jacky, give me paper and I will write.' I gave him pencil and paper, and he tried to write, and he then fell back and died, and I caught him in my arms and held him; and I then turned round myself and cried. I was crying a good while until I got well; that was about an hour, and then I buried him.

"I digged up the ground with a tomahawk, and covered him over with logs and grass, and my shirt and trousers. That night I left him near dark. I would go through the scrub and the blacks threw spears at me; a great many; and I went back into the scrub. Then I went down the creek which runs into Escape River, and I walked along the water in the creek, very easy, with my head only above water, to avoid the blacks, and get out of their way. In this way I went half-a-mile. Then I got out of the creek, and got clear of them, and walked all night nearly, and slept in the bush without a fire."

At the southern entrance of Albany Pass, one of the most picturesque spots of the east coast of Australia, the schooner "Ariel" lay at anchor, awaiting, day after day, some signal to indicate the arrival of the expected Kennedy. One day the look-out man announced that there was an aboriginal on the mainland making urgent signals to the schooner. There was nothing unusual in this, for during the delay and tedious waiting, the blacks had constantly been seen making gestures on the shore. An examination through the glass, however, showed the people on the "Ariel" that this blackfellow was making such vehement and persistent signals that it was thought worth while to send the boat in to investigate affairs.

No wonder the poor fellow's signals were urgent and vehement; he was Jacky-Jacky, who, thirteen days after Kennedy's death, by devious twistings and windings, occasionally climbing a tree in the hope to catch a glimpse of the schooner, and existing on roots and vermin, had at last reached the goal. But when he stood prominently on the shore to signal to the schooner, his relentless pursuers sighted him, and his frantic signs were for rescue from imminent peril. The boat's crew fortunately recognised the emergency, and a smart race ensued between them and the natives. The rescuers won, and Jacky-Jacky was saved to tell his melancholy story.

There was no time lost on board the "Ariel." There were three men who might be still alive at Shelburne Bay, and eight more starving at Weymouth Bay. Kennedy was dead; their duty, and urgent duty it was, lay with the living. At once the schooner commenced to beat down the coast, and at Shelburne Bay they landed but failed to find the camp. But they seized a native canoe which bore sufficient evidence that the men had been murdered. Clearly time must not be wasted in inflicting punishment; according to Jacky's account, the men at Weymouth Bay were absolutely starving, if they had not already succumbed to famine.

After their leader had left Weymouth, Carron had shifted the camp on to the nearest hill, as it was more open and less exposed to the treacherous attacks of the natives. A flagstaff was erected on the crest, in view of the Bay. Then the party had only to sit down and await the coming of the grim shadow following them through the jungle to strike them with the death chill. They had two skeletons of horses and two gaunt dogs, and a tiny remnant of flour. The men gave themselves up to moody despondency. "Wearied out by long endurance of trials that would have shaken the courage and tried the fortitude of the strongest," says Carron in his diary, "a sort of sluggish indifference prevailed that prevented the development of those active energies which were necessary to support us in our present critical position."

One of the two horses was killed, and its scanty flesh, cut into strips, was dried in the sun and smoke. This, the most repellant, sapless food to be found in the world, had been their diet for some time. Douglas was the first to die. The survivors were still strong enough to give him burial. In a few days Taylor followed him and was interred by his side. The blacks threatened them continually, though at times they would lay down their arms and bring pieces of fish and turtle into the camp; but this only the better to spy out their weakness. Carpenter was the next to succumb, and on the 1st of December they were doomed to drink their bitterest cup to the dregs. They had killed the remaining horse, but the monsoonal rains descended, and in the steamy atmosphere the meat turned putrid. Torn with anxiety, Carron was dejectedly mounting the look-out to the flagstaff when he caught sight of a vessel beating into the Bay. The sudden change from despair to relief was overwhelming. Kennedy must have reached Port Albany, and had doubtless sent the "Bramble" to rescue them. With eager, tremulous hands he hoisted a pre-arranged signal to warn them against the blacks. Darkness fell and they kept a fire burning, and fired off rockets, and when daylight came and a boat was lowered from the schooner, they felt no misgivings. Time passed, and Carron again ascended the look-out. What he saw nearly blasted his eyesight. The schooner was standing out to sea; he was just in time to see her round the point and disappear.

They strove to persuade themselves that it was not the "Bramble," a relief schooner that was supposed to cruise along the coast. But it assuredly had been the "Bramble," and her men had not seen the signals against the gloomy background of scrub and hills. They knew nothing of Kennedy's death, nor of Carron's plight. The agony of this disappointment must have been more bitter than death. Mitchell was the next to die, and the survivors were too weak to give him burial. Then Niblett and Wall departed, but on the last day of the year relief came to the remaining two.

Some natives suddenly brought Carron a dirty note, to say that help was coming, and he saw by their gestures that there was a vessel in the bay. He scribbled a note in reply, but they refused to take it, and began to crowd into the camp and handle their weapons. They were not going to be baulked of their prey. At the very moment when they were poising their spears, the relief party arrived. Four brave men—Captain Dobson of the "Ariel," Dr. Vallack, Barrett a sailor, and the eager Jacky-Jacky—had forced their way through mangroves and hostile threatening natives to snatch them from their doom.

Nothing could be carried away but the two famished men, and they were helped down to the boat without coming into active hostilities. Thus ended the most disastrous expedition in Australian annals. Kennedy's body was never recovered, nor was the fate of the men at Shelburne Bay revealed. The bodies at Weymouth Bay were re-buried on Albany Island, and a tablet was erected in memory of Kennedy, in St. James's Church, Sydney.

Chapter X.

LATER EXPLORATION IN THE NORTH-EAST.

(i.)—Walker in Search of Burke and Wills.

Frederick Walker commenced his bush career as a pioneer squatter in the districts of Southern Queensland, but afterwards made his residence near the centre, where he joined the Native Police. He had long bush experience, was a firm believer in the training of the natives in quasi-military duty, and had taken a prominent part in the formation of the Queensland Native Police. On this relief expedition, the party was composed almost entirely of Native Police troopers under his leadership.

On receiving his commission, he pushed rapidly out to the Barcoo, and, near the Thomson River, came upon another tree marked L. This might have been made by Leichhardt. He ascended the main watershed, and crossed it coming down on to the head of the Flinders River. Here he experienced many hindrances arising from the rough basaltic nature of the country that borders the northern head-waters of that river. When he finally debouched upon the wide western plains, he crossed the Flinders, without recognising it as the main branch, in the search for which he went on northward. Approaching the Gulf of Carpentaria, he had several encounters with the aboriginals. As he neared the coast, the bend of the Flinders brought that river again across his route, and it was then that he came on some camel tracks, which assured him that the missing party, the object of his search, had at any rate reached the Gulf safely. On his outward way Walker may be said to have pursued a course parallel with that of the Flinders, a little further to the northward.

He pushed on to the Albert River, to replenish his provisions at the depot provided for the use of the various relief parties. He arrived there safely, after having had two more skirmishes with the blacks on the way. He reported the finding of the camel tracks, and having come to the conclusion that Burke and Wills had probably made for the Queensland settlements, he decided to follow them thither. He traced out a tributary of the Flinders, the Saxby, on his homeward route, but saw no more of the camel tracks, and finally crossed the watershed on to the rough basaltic country at the head of the Burdekin. Here his horses suffered so severely from the rugged nature of the country, that by the time they reached Strathalbyn, a station on the lower Burdekin, the whole of the party were well-nigh horseless, as well as almost out of provisions.

Walker was afterwards engaged by the Queensland Government to mark out a course for a telegraph line between Rockingham Bay and the mouth of the Norman River in Carpentaria. This work he carried out successfully; but when at the Gulf, he was attacked by the prevalent malarial fever, and died there.

(ii.)—BURDEKIN AND CAPE YORK EXPEDITIONS.

The main portion of eastern Australia was now fairly well-known; it had been crossed from south to north, and from east to west, and it was only the elongated spur of the Cape York peninsula that stood in urgent need of detailed exploration.

Amongst what may be called the minor pastoral expeditions of that period, was one conducted by G. E. Dalrymple, who penetrated the coastal country north of Rockhampton as far north as the Burdekin. In 1859 he followed that river down to the sea, and found that the mouth had been located further to the south than was really the case. His party then struck inland, examined the head of that river, and found the Valley of Lagoons. The following year another party, consisting of Messrs. Cunningham, Somer, and three others, explored the tributaries of the Upper Burdekin, and opened up several good tracts of pastoral country. The permanent running stream which flows through a rugged wall of basalt into an ana-branch of the Burdekin, was first noticed by this party, and called Fletcher's Creek.

Frank and Alec Jardine jointly led up the Cape York Peninsula an expedition that in its hardships and dangers emulated that of Kennedy's, but fortunately without a tragic ending. The year 1863 was one of great activity in the northern part of eastern Australia. At Cape York, the Imperial Government had, on the recommendation of Sir George Bowen, the first Governor of Queensland, decided to form a settlement. John Jardine, the police magistrate of the central town of Rockhampton, was selected to take charge, and a detachment of marines was sent out to be stationed there. Somerset, the new settlement, was formed on the Albany Pass, opposite to the island of the same name.

Frank L. Jardine.

Jardine was to proceed by sea to his new sphere of office, but, anticipating the want of fresh meat at the proposed station, he entered into an arrangement with the Government whereby his two sons were to take a small herd of cattle thither overland, and on the way make careful observations of the land through which they were to pass. Somerset was situated near the scene of Kennedy's death, and knowing what tremendous difficulties that explorer had met with on the eastern shore, it was decided that the expedition should attempt to follow the western shore through the unknown country that faced the Gulf of Carpentaria. Both the Jardine brothers were quite young men at the time when they

started on their exceedingly adventurous trip, which
combined cattle-droving with exploration; Frank, the
accepted leader, being only twenty-two years old, and his
brother Alexander but twenty. Their father had come
from Applegarth, in Dumfriesshire; they had both been
born near Sydney, and had been educated by private
tutors and at the Sydney Grammar School.

They took with them A. J. Richardson, a surveyor
sent by the Government, Scrutton, Binney, Cowderoy,
and four natives. The stock consisted of forty-two
horses and two hundred and fifty head of cattle. The
cheerful acceptance of this hazardous enterprise by these
youths was a fine indication of adventurous spirit, and
reflects great credit on their courage and the courage
of the native-born. The fate of the last explorer who
dared to face the perils of the Peninsula would have
deterred any but the boldest from taking up his task.

Before the final start from Carpentaria Downs, then
the furthest station to the north, supposed to be situated
on Leichhardt's Lynd River, Alec Jardine made a trip
ahead in order to secure knowledge of an available
road for the cattle, and save delay in the earlier stages
of the main journey. On this preliminary observational
excursion, he followed the presumed Lynd down for
nearly 180 miles, until he was convinced that neither in
appearance, direction, nor position did it correspond with
the river described by Leichhardt. On the subsequent
journey with the cattle, this conviction was found to be
in accordance with fact, for the stream was then proved
to be a tributary of the Gilbert, now known as the
Einnesleigh.

On the 11th of October the final start was made, and
the party commenced a journey seldom equalled in
Australia for peril and adventure. The head of the
Einnesleigh was amongst rough ranges, and on the 22nd
of the month they halted the cattle while they conducted
another search for the invisible Lynd. They found other
good-sized creeks, but no Lynd, nor did they ever see it.
They afterwards found that, owing to an error in the

map they had with them, the Lynd was placed 30 miles out of position. A misfortune happened at the outset of their expedition. In the morning a large number of horses were missing. Leaving some of the party to stay behind and look for them, the two brothers and the remainder went on with the cattle. On the second day they arrived at a large creek, without having been overtaken by the party with the missing horses and the pack-horses. After an anxious day spent in waiting, Alec Jardine started back to find out the cause of the delay. He met the missing party, who were bringing bad news with them. Through carelessness in allowing the grass round the camp to catch fire, half of their rations and nearly the whole of their equipment had been burnt. In addition, one of the most valuable of their horses had been poisoned. This terrible misfortune, coming at such an early stage of their journey when they had all the unknown country ahead of them, seriously imperilled the success of their undertaking. But there was nothing to do but to bear it with what equanimity they could muster.

Alec W. Jardine.

The Cape York natives now seemed to rejoice that they had another party of white men to dog to death. Once about twenty of them appeared about sundown and boldly attacked the camp with showers of spears. Two days afterwards, they surprised the younger Jardine when alone, and he had to fight hard for his life. The creek they had been following down led them on to the Staaten

River, where the blacks succeeded in stampeding their horses, and it was days before some of them were recovered.

On the 5th of December, they left this ill-omened river, and steered due north. Bad luck still haunted them; tortured by flies, mosquitoes, and sand-flies, their horses scattered and rambled incessantly. While the brothers were absent, searching one day for the horses, the party at the camp allowed the solitary mule to stray away with its pack on. The mule was never found again, and it carried with it, in its pack, some of their most necessary articles, reducing them nearly to the same state of deprivation as their determined enemies, the aboriginals. Two more horses went mad, through drinking salt water; one died, and the other was so ill that he had to be abandoned. On the 13th of December they reached the Mitchell River, not without having had another hot battle with the blacks, who followed them day after day, watching for every opportunity and displaying the same relentless hostility that they had formerly shown to Kennedy. Whilst the party were on the Mitchell, the natives mustered in force and fell upon the explorers with the greatest determination. After a severe contest, in which heavy loss had been inflicted upon the savages, they sullenly and reluctantly retired. From what was afterwards gathered from the semi-civilised natives about Somerset, these tribes followed the Jardines for nearly 400 miles. This perseverance and inappeasable enmity had been equalled before only by the Darling natives. It can be imagined how these incessant attacks, combined with the harassing nature of the country, gave the party all they could do to hold their own, and but for the prompt and plucky manner in which the attacks were met, not one of them would have survived.

After crossing the Mitchell, steering north, they got into poor country, thinly-grassed and badly-watered, with the natives still hanging on their flanks. On the 28th of December, the blacks began to harass the horses, and another hard struggle took place. Storms of rain now set in, and they had to travel through dismal tea-tree flats, with the constant expectation of being caught by a flood in the low-lying country.

In January, they had a gleam of hope. On the 5th they came to a well-grassed valley, with a fine river running through it, which they named the Archer. On the 9th they crossed another river, which they supposed to be the one named the Coen on the seaward side. But once across this river, troubles gathered thick again; the rain poured down constantly, the country became so boggy that they could scarcely travel, and to crown all their misfortunes, two horses were drowned when crossing the Batavia, and six others were poisoned and died there.

Fate seemed now to have done her worst, and the explorers faced the future manfully. Burying all that they could dispense with, they packed all their remaining horses and started resolutely to finish the journey on foot. On the 14th two more of their horses died, and the blacks once more came up behind to reconnoitre. As may be imagined, the whites were not in a patient humour, and this last skirmish was brief and severe.

On the 17th two more horses died from the effects of the poison plant. Fifteen only were left out of the forty-two with which they had started. They were now approaching the narrow point of the Cape, and found themselves on a dreary waste of barren country whereon only heath grew, and which was intersected with boggy creeks.

On the 10th of January, they caught a glimpse of the sea from the top of a tree, and on the 20th they were in full view of it. As they went on, they were entangled in the same kind of scrub that baffled Kennedy, and at last on the 29th, after some days of scrub-cutting, it was determined to halt the cattle, whilst the brothers should push on to Somerset in the endeavour to find a more practicable track. In the tangled, scrubby country through which they had passed, it had been difficult to form a true conception of the distance, and their estimate of twenty miles for the distance separating them from the settlement was much too short.

On the 30th of January, the two Jardines and their most trusted black boy, Eulah, started to find the settlement. For a time they were hemmed in by a bend of what they took to be the Escape River, but on getting clear of it, they were surprised to come to another large and swollen river, which apparently ran into the Gulf. This forced them to return. After a few days' rest, they made a second vain attempt. Hemmed in by impassable morasses and impenetrable thickets, in some places they were cut off from approaching even the river, by formidable belts of mangroves. In fact, the Jardine River, as it is now called, heads almost from the eastern shore, from Pudding Pan Hill in fact, Kennedy's fatal camp. It overlaps the Escape River, and after many devious windings and twistings, flows across the Cape, out on to the Gulf shore.

It was not until the end of February that, on the subsidence of some of the flooded creeks, the brothers made a successful effort, and got into somewhat better travelling country. The next morning they came across some blacks who were eager to be on good terms, and hailed them to their surprise with shouts of "Franco; Allico; Tumbacco". These cries had been taught them by Mr. Jardine, who was getting anxious because of his sons' delay, and had done all he could think of to help them. He had cut a marked tree line, almost from sea to sea; and coached the local natives up in a few English words, so as to be recognised as friends. This last device succeeded admirably. From these newcomers, they selected three as guides, and the following day reached the settlement.

The rest of the party and the stock were soon brought into Somerset, where a cattle-station was formed. When we look back at the difficulties that beset the path of this expedition, and the unforeseen disasters that befel them, one cannot help feeling the greatest admiration for the leaders and their conduct. In spite of the numberless treacherous attacks of the blacks to which they had been subjected, not a member of the band had been lost. They had fought their way through the same species of danger that had environed the unfortunate Kennedy, and had all lived to tell the tale. The Royal Geographical Society rewarded the labours of the two brothers by electing them Fellows of the Society, and by awarding them the Murchison medal.

Frank Jardine was for some period Government Resident at Thursday Island, whither the settlement has been removed; but of late he has resided at his own station at Somerset, and engaged in pearl-shelling. Alec entered the Queensland civil service, as Roads Engineer, and in that capacity did much important work in the construction of the roads of that State. In 1871-1872, he designed and constructed the road and railway-bridge over the Dawson River, and in 1890 he became Engineer-in-Chief for Harbours and Rivers.

But the scrubby and hilly nature of the country on Cape York militated against its speedy settlement, and it needed the lure of gold to induce men to risk their lives in a land with such hostile inhabitants. In 1872 the Queensland Government decided upon another exploration of the neck of land that forms the northern-most point of Australia. More than eight years had elapsed since the Jardines had made their dashing journey; but their report, coupled with Kennedy's fate, did not offer much temptation to follow up their footsteps. There was, however, a tract of country near the base of the Peninsula still comparatively unknown; and a party was organised and placed under the leadership of William Hann. Hann was a native of Wiltshire, who had come out to the south of Victoria with his parents at an early age. He was afterwards one of the pioneer squatters of the Burdekin, in which river his father was drowned. The object of the trip was to examine the country as far as the 14th parallel S., with a special view to its mineral resources. The discovery of gold having extended so far north in Queensland had raised a hope that its existence would be traced along the promontory. Hann had with him Taylor as geologist, and Dr. Tate as botanist, the latter being a survivor of the melancholy "Maria" expedition to New Guinea. Apparently his ardour for exploration had not been cooled by the narrow escape he had then experienced.

The party left Fossilbrook station on the creek of the same name, a tributary of the Lynd, north of the initial point of the Jardine expedition. Crossing much rugged and broken country, they found two rivers running into the Mitchell, and named them the Tate and the Walsh. From the Walsh, the party proceeded to the upper course of the Mitchell, and crossing it, struck a creek, marked on Kennedy's map as "creek ninety yards wide." This was named the Palmer, and here Warner, the surveyor, found traces of gold. A further examination of the river resulted in likely-looking results being obtained; and the discovery is now a matter of history, the world-wide "Palmer rush" to north Queensland being the result in 1874.

On the 1st of September, Hann reached his northern limit, and the next day commenced the ascent of the range dividing the eastern and western waters. A few days afterwards, he sighted the Pacific at Princess Charlotte Bay. From this point the party returned south, and came to a large river which he called the Normanby, where a slight skirmish with the natives occurred, the blacks having hitherto been on friendly terms. While the men were collecting the horses in the morning, the natives attempted to cut them off, each native having a bundle of spears. A few shots at a long distance were sufficient to disperse them, and the affair ended without bloodshed.

On the 21st of September, Hann crossed the historical Endeavour River, and upon a small creek running into this inlet, he lost one of his horses from poison. Below the Endeavour, the party encountered similar difficulties to those that dogged poor Kennedy's footsteps—impenetrable scrub and steep ravines. This went on for some days, and an attempt to reach the seashore involved them in a perfect sea of scrub, and necessitated the final conclusion that advance by white men and horses was impossible. Hann had reluctantly to make up his mind to return by the Gulf Coast, and abandon the unexplored ground to the south of him.

After many entanglements in the ranges, and confusion arising from the tortuous courses of the rivers, the watershed was at last crossed, and on the 28th of October they camped once more on the Palmer, whence they safely returned along their outward course.

The gold discoveries on the Palmer, and the rush caused thereby, coming soon after this expedition, led to a great deal of minor exploration done under the guise of prospecting; and it is greatly to the work of prospectors for gold that much of the knowledge of the petty details of the geographical features of Australia is due. To the courage and endurance of this class of settler, Australia owes a great debt, but their labours are unrecorded and often forgotten.

Part II.

CENTRAL AUSTRALIA.

Statue of John McDouall Stuart, in the Lands Office, Sydney.

Chapter XI.

EDWARD JOHN EYRE.

(i.)—Settlement of Adelaide and the Overlanders.

The exploration of the centre of the continent was long retarded by the difficult nature of the country—by its aridity, its few continuously-watered rivers, and the supposed horse-shoe shape of Lake Torrens, which thrust its vast shallow morass across the path of the daring explorers making north.

For most of us of the present day, to whom Lake Torrens is but a geographical feature, it is hard to imagine the sense of awe it inspired in the breasts of the South Australian settlers, who appeared to be cut off completely from the north by its gloomy and forbidding environs of salt and barrenness.

In 1836, Colonel Light surveyed the shores of St. Vincent's Gulf, and selected the site of the city of Adelaide. Governor Hindmarsh and a company of emigrants arrived soon afterwards, and the Province of South Australia was proclaimed.

The very promising discoveries made to the south of the Murray by Major Mitchell soon induced an invasion of adventurous pastoralists bringing their stock from the settled parts of New South Wales.

Charles Bonney led the way across to the Port Phillip settlement in 1837 with sheep. G. H. Ebden accompanied him, and they were shortly followed by many more: Hamilton, Gardiner, Langbourne, and others, whose names are well-known in Australian history as "the first Overlanders." Very shortly this overlanding of stock was extended to the newly-founded city of Adelaide, Charles Bonney and Joseph Hawdon being the first drovers on this long journey. Their Adelaide journey was in fact an exploration trip, and an important one, as they followed the bank of the Murray below its junction with the Darling; this part of the river having been followed down before only by Sturt, and then only by water.

It was in January, 1838, that Hawdon and Bonney left Mitchell's crossing at the Goulburn River with cattle as pioneers on the overland route to Adelaide. Unknown to them they were closely followed by E. J. Eyre, with another mob of cattle. Eyre, as we shall afterwards see, was thrown out of the race through trying to make a short cut to avoid the sweeping bend of the river. Bonney and Hawdon crossed the Murray above the junction of the Darling, and in places found the bed of the latter river dry. The natives, strange to say, were quite friendly; perhaps they had taken to heart the lesson Mitchell had read them. But their amiable demeanour did not last long. Bonney and Hawdon were almost the last overlanding party to proceed unmolested. Within a comparatively short time afterwards, an incessant war began to be waged between the blacks and every Overlander who passed down the Murray. It ended only with the sanguinary battle of the Rufus. More fortunate than Sturt, Hawdon and Bonny were able to cut off many of the wearisome bends that had so fatigued Sturt's crew. Sturt had had to follow every turn and curve, whilst the Overlanders avoided the bends of the Murray by following the native paths, which spared them in some cases a journey of one or two days. It was while following a native path that they discovered and named Lake Bonney. At last they sighted the Mount Lofty ranges, and after some difficulty in getting through some rough mallee-covered country, arrived at Adelaide, and gladdened the residents with the prospect of roast beef. "Up to this time," says Bonney in his diary, "they had been living almost exclusively on kangaroo flesh." Eyre, whose name was afterwards so closely allied with a famous story of thirst and hardship, narrowly escaped with his life during his overlanding trip.

It was owing to a very natural mistake that Eyre was led astray. He intended to try a straighter and shorter route than the one round the Murray, and for a time got on very well, but coming across a tract of dry country across which he could not take the cattle, he determined to follow Mitchell's Wimmera River to the north, naturally thinking that it would lead him easily to the Murray, and would probably prove to be identical with the Lindsay, as marked on Sturt's chart. From Mitchell's furthest point, he traced it a considerable distance to the north-west, and at last found its termination in a large swampy lake, which he called after the first Governor of South Australia, Lake Hindmarsh. From this lake he could find no outlet, so taking with him two men, he made an attempt to push through to the Murray, leaving his cattle to await him. He found the country covered with an almost impenetrable mallee scrub, and as there was neither grass nor water for the horses, he was forced to retreat. He reached his camp after a weary struggle on foot, the horses having died from thirst. Eyre was then compelled to return and gain the bank of the Murray by the nearest available route. The bitter disappointment of the trip was, that when forced to retreat by the inhospitable nature of the country, he was but twenty-five miles from the river.

Bonney, however, on another occasion, took a mob of cattle from the Goulburn River to Adelaide in almost a direct line. In February, 1839, he left the Goulburn and steered a course for the Grampian Mountains, where he struck the Wannon, and followed it down to the Glenelg. Here he came upon one of the Henty stations, and was strongly advised not to persist in his attempt. Captain Hart, who had been examining the country with the same purpose in view as Bonney's, stated that it would be impossible to take cattle through, and turned back with his own to follow the old route round the Murray bend. But Bonney was not to be daunted, and resolutely pushed on west of the Glenelg. He discovered and named Lake Hawdon, and also named two mountains, Mount

Muirhead and Mount Benson. But at Lacepede Bay his
most serious troubles commenced. The party had pushed
on steadily to within forty miles of Lake Alexandrina
when, in the middle of a sandy desert, the working
bullocks failed. Bonney divided his party, and sending
some of the men back to take the workers to a brackish
pool which they had passed, he himself with the stockmen
and two black boys, made a desperate effort to reach the
Lake with the main mob. For two days they pushed
steadily on, travelling day and night, until men and beasts
were alike at their last gasp. Bonney then tried a
desperate expedient:—"I then determined," he says, "as
a last resource, to kill a calf and use the blood to assuage
our thirst. This was done, and though the blood did not
allay the pangs of thirst to any great extent, it restored
our strength very much."

The exhausted men then lay down to rest; but whilst
they slept, their thirsty beasts scented a faint smell of
damp earth on a wandering puff of wind, and stampeded
off to windward. Too weak to follow on at once, the men,
after an hour or two, staggered after them and tracked
them to a half-dry swamp, which still retained a little
mud and water. It was brackish, but palatable enough
for men in their exhausted condition, and saved the lives
of all. After some trouble in crossing the Murray, they
reached Adelaide in safety with the stock.

When the news of their arrival reached Port Phillip,
many other Overlanders were encouraged by Bonney's
example to try the shorter route, and the trade in
shipping cattle across the straits from Tasmania almost
ceased.

Bonney had been born at Sandon, near Stafford,
and educated at the Grammar School, Rugby. He had
come out to Sydney in 1834, as clerk to Sir William
Westbrooks Burton; but the love of adventure prevailed
over his other inclinations, and in 1837, he joined Ebden
in squatting pursuits, and eventually distinguished
himself as one of the leading Overlanders. He
subsequently settled in South Australia. From 1842 to
1857 he was Commissioner for Crown Lands, and he

afterwards served the State as manager for railways, and in other capacities. Subsequently he returned to Sydney, where he died.

(ii.)—Eyre's Chief Journeys.

Edward John Eyre was the son of the Rev. Anthony Eyre, vicar of Hornsea and Long Riston, Yorkshire, and was born on August 14th, 1815. He was educated at Louth and Sedburgh Grammar Schools. He came to Australia in 1833, and immediately engaged in squatting pursuits, his enterprising spirit constantly leading him beyond the pale of civilization, where his natural love for exploration rapidly increased. His fortunes as an Overlander have already been noticed. On the 5th August, 1839, he left Port Lincoln, on the western shore of Spencer's Gulf, meaning to penetrate as far as he could to the westward.

Edward John Eyre.

Some time before he had made an expedition to the north of Adelaide as far as Mount Arden, a striking elevation to the N.N.E. of Spencer's Gulf. He had ascended this mount, and from the summit seen a depression which he took to be a lake with a dry bed. This lake afterwards played an important part in the history of South Australian settlement under the name of Lake Torrens.

Eyre's party on his westward trip consisted of an overseer, three men, and two natives. Twenty days after leaving Port Lincoln, they arrived at Streaky Bay, not having crossed a single stream, rivulet, or chain of ponds the whole distance of nearly three hundred miles. Three small springs only had been found, and the country was covered with the gloomy mallee and tea-tree scrub. Westward of Streaky Bay the country was still found to be scrubby; so Eyre formed a camp, and taking only a black boy with him, he forced a stubborn way onward, until he was within nearly fifty miles of the western border of South Australia. To all appearance the country was slightly more elevated than the level scrubby flats he had been traversing, but there was neither grass nor water, and an immediate return became necessary. Before he got back to Streaky Bay camp, he nearly lost three of his horses.

Leaving Streaky Bay again, he went east of north to the head of Spencer's Gulf, finding the country on this route a little better, but still devoid of water, the party getting through, thanks only to a timely rainfall. On the 29th of September, he came to his old camp at Mount Arden, where he wrote:—

"It was evident that what I had taken on my last journey to be the bed of a dry lake now contained water, and was of a considerable size; but as my time was very limited, and the lake at a great distance, I had to forego my wish to visit it. I have, however, no doubt of its being salt, from the nature of the country, and the fact of finding the water very salt in one of the creeks draining into it from the hills. Beyond this lake (which I distinguished with the name of Colonel Torrens) to the westward was a low, flat-topped range, extending north-westerly, as far as I could see."

From this point Eyre returned, pursuing his former homeward route.

The main objects that now attracted the attention of the colonists of South Australia were (1) discovery to the northward, regarding both the extent of Lake Torrens and the nature of the interior; and (2) the possibility of the existence of a stock route to the Swan River settlement. Eyre, however, after his late experience, was convinced that the overlanding of stock around the head of the Great Bight was impracticable. The country was too sterile, and the absence of water-courses rendered the idea hopeless. For immediate practical results, beneficial

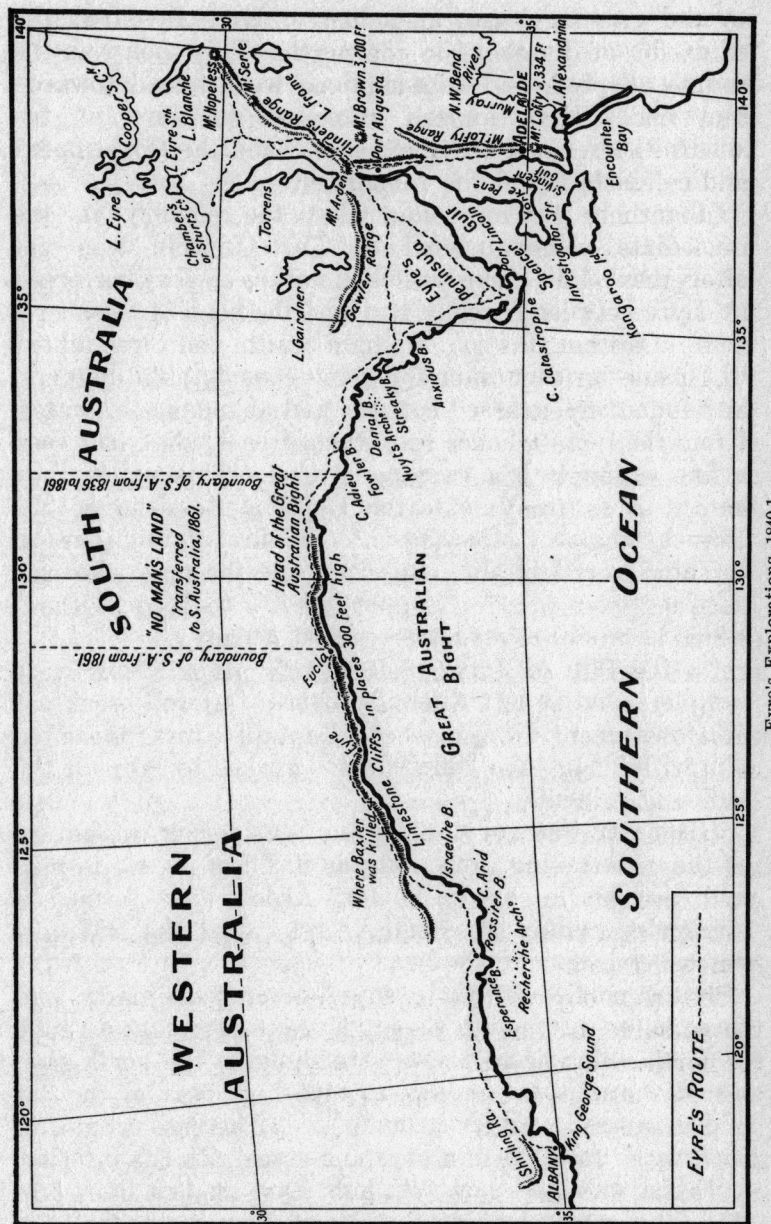

Eyre's Explorations, 1840-1.

to the growing pastoral industry, Eyre favoured the extension of discovery to the north. This then was the course adopted, and subscriptions were raised towards that end. Eyre himself provided one-third of the needful horses and other expenses; and the Government and colonists found the remainder.

Meantime it was found that the country in the immediate neighbourhood of Port Lincoln was not altogether of the same wretched nature as that traversed by Eyre between Streaky Bay and the head of Spencer's Gulf. Captain Hawson, William Smith, and three others had made an excursion for some considerable distance, and found well-grassed country and abundance of water. From the point whence they turned back, they had seen a fine valley with a running stream. This valley they named Rossitur Vale, after Captain Rossitur of the French whaler "Mississippi," the first foreign vessel to enter Port Lincoln. Rossitur was the man who was destined later to afford opportune aid to Eyre, without which he would never have reached Albany.

On the 18th of June, 1840, Eyre's preparations were complete, and he left Adelaide after a farewell breakfast at Government House, where Captain Sturt presented him with a flag—the Union Jack—worked by some of the ladies of Adelaide.

His party was not a large one considering the nature of the undertaking, consisting as it did of six white men and two black boys. At Mt. Arden they formed a stationary camp. A small vessel called the "Waterwitch," was sent to the head of Spencer's Gulf with the heaviest portion of their supplies, and the party had three horse drays with them. Eyre trusted that a range of hills, which he had seen stretching to the north-east, would continue far enough to take him clear of the flat and depressed country around Lake Torrens—would, in fact, as he says, form a stepping-stone into the interior.

Taking one black boy with him, Eyre made a short trip to Lake Torrens, leaving the rest of the party to land the stores from the "Waterwitch." He found the bed of the lake coated with a crust of salt, pure white, and glistening brilliantly in the sunshine. It yielded to the footstep, and below was soft mud, which rapidly grew so boggy as to stop their progress. In fact they had to return to the shore without being able to ascertain whether there was any water on the surface or not. At this point the lake appeared to be about fifteen or twenty miles across, having high land bounding it on the distant west.

There seemed no chance of crossing the lake; and following its shore to the north was impossible. There was neither grass nor water; the very rainwater turned salt after lying a short time on the saline soil. The only chance of success appeared to be to keep close to the north-eastern range, which Eyre named the Flinders Range, trusting to its broken gullies to supply them with some scanty grass and rainwater.

It was a cheerless outlook. On one side was an impassable lake of combined mud and salt; on the other a desert of bare and barren plains; whilst their onward path was along a range of inhospitable rocks.

"The very stones, lying upon the hills," says Eyre, "looked like scorched and withered scoria of a volcanic region, and even the natives, judging from the specimen I had seen to-day, partook of the general misery and wretchedness of the place."

He directed his course to the most distant point of the Flinders Range, but when he arrived there, he was obliged to christen it Mount Deception, as his hope of finding water there was disappointed. Subsisting as well as they could on rain puddles on the plains, Eyre and his boy searched about for some time and at last found a permanent-looking hole in a small creek. They then returned to the main party. Having concealed the supplies landed from the cutter, Eyre sent the vessel back to Adelaide with despatches, and moved the whole of the men out to the pool of water that he had just found. From this vantage point he made various scouting trips with the black boy, both to the eastward and westward of north. The 2nd of September found him on the summit of an elevation which he appropriately named Mount Hopeless, gazing at the salt lake that he now thought hemmed him in on three sides, even to the eastward. There was no prospect visible of crossing the lake, which seemed persistently to defy him, meeting him at every attempt with a barrier of stagnant mud. There was nothing for it but to leave the interior unvisited by this route, and to return to Mount Arden.

He divided his party, sending Baxter, the overseer, with most of the men and stores straight across to Streaky Bay, where he had formerly made a camp, while, with the remainder, he made his way to Port Lincoln. Having abandoned his intention to penetrate to the interior on a northern course, he now determined to push out westward, to King George's Sound, finding, perhaps, on the way across, some inducement that would lead him north.

At Port Lincoln he could not obtain the extra supplies he wanted without sending to Adelaide; it was therefore the 24th of October when he finally started for Streaky Bay. He found that Baxter had arrived there safely, and was anxiously awaiting him.

He now camped for many weeks at Fowler's Bay, which was as far as the cutter they now had, the "Hero," could act as convoy, her charter not extending beyond South Australian waters. The "Waterwitch" having sprung a leak, the "Hero" had taken her place. During the time that they remained there, Eyre made many journeys ahead to estimate his chances of getting across the dry and barren country intervening between him and the Sound, but the outlook was disheartening. He met some natives, who all assured him that there was no water ahead; nor could he find any but some brackish water obtained by digging in some sandhills. Worse than all, he sacrificed three of his best horses during these fruitless attempts.

On the 25th of January, the "Hero" arrived with the oats and bran he had sent back for. So poverty-stricken was the country that Eyre, in the circumstances, resolved to send back nearly the whole of his expedition by the vessel, and then, with only a small party, to push through to King George's Sound or perish in the attempt.

Baffled successively to the north and to the west, Eyre had been put upon his mettle, and he could not endure the thought of returning to Adelaide a beaten man.

On the 31st of January the cutter departed, and Eyre, Baxter, and three native boys, one of whom had come by the vessel on her last trip, were left alone to face the eight hundred miles of desert solitude before them. Some time was spent in making their final preparations, but on the 24th of February they had actually begun their journey when, to their astonishment, they heard two shots fired at sea. Thinking that a whaler had put in to the bay, Eyre turned back, but found the "Hero" again in port with an urgent request from Adelaide to abandon his desperate project, and return in the vessel. Upon a man of Eyre's temperament, this recall could have only one effect, that of strengthening his resolve to proceed westward at all hazards. He did not emulate Cortez by burning his ship behind him, but he none the less effectually deprived himself of means of retreat by dismissing the little "Hero."

It was at the close of a hot summer when Eyre started, and the nature of the sandy soil, combined with the low prickly scrub, soon began to hamper their progress and render the lack of water especially severe. On one side of them, flanking their line of march, were the cliffs of the Great Bight, against which thundered the ever-restless southern rollers; on the other there stretched a limitless expanse of dark, gloomy scrub. Their only hope of relief was the faint chance of striking some native path which might lead them to an infrequent soakage-spring. Even in these depressing circumstances, Eyre seems to have found time to express his admiration of Nature as she then revealed herself to him:—

"Distressing and fatal as the continuance of these cliffs might prove to us, there was a grandeur and sublimity in their appearance that was most imposing, and which struck me with admiration. Stretching out before us in lofty, unbroken outline, they presented the singular and romantic appearance of massy battlements of masonry, supported by huge buttresses, glittering in the morning sun which had now risen upon them, and made the scene beautiful even amidst the dangers and anxieties of our situation.''

Five days of slow, dragging toil passed, until, with the horses at their last gasp, and the men baked and parched, they found relief in some native wells amongst the sandhills, at a point where the cliffs receded from the sea.

After resting for some days at this camp, Eyre, misled by a report he had obtained from the natives, again moved forward, taking with him but a small supply of water. When he had discovered the blunder, he had gone forty miles, and over this weary distance the horses had to return. It was one of those mishaps that helped so much to wear out his unfortunate animals.

Trouble after trouble now added itself to the burden of the explorers. Another five days had passed without water, and their only hopes rested upon some sandhills ahead, seen from the sea by Flinders, and marked by him upon his chart. Retreat was impossible, and with their horses failing one after another, they toiled on, desperate and well-nigh hopeless. Eyre's anxiety was increased by Baxter's growing despondency and pessimistic view of the issue of their enterprise. They were now travelling along the sea beach, firm and hard, and ominously marked with wreckage. Their last drop of water had been consumed, and that morning they had been collecting dew from the bushes with a sponge, as a last resource. When they reached the sand-dunes, they were almost too weak to search for a likely place to dig for water; but making a final effort, they discovered a patch whence, at six feet, they obtained a supply of water.

It was now that Eyre approached the grand crisis of his adventurous journey. According to the chart compiled by Flinders, he had another long succession of cliffs to encounter, and he knew that where these cliffs came in and sternly fronted the ocean, he need hope for no relief. Should this space be happily surmounted by a desperate effort, he hoped to reach a kindlier country. Disaffection appeared in his small camp. Baxter was always suggesting and even urging a return. Perhaps some shadow of his tragic fate overhung his spirit. The native boys were ripe for desertion, and two of them did desert, only to return in a few days, starving, and apparently repentant. Better for Eyre had they gone altogether. Amid such discouraging surroundings did Eyre commence his last struggle with the cliffs of the Great Bight.

The party had been tantalised by threatening clouds, which never broke in rain. When on the third day they gathered once more, black and lowering, Baxter urged Eyre to camp that night instead of pushing on, as rain seemed certain, and the rock holes by which they were then passing were well adapted to catch the slightest shower. Eyre consented, against his better judgment. It was necessary to watch the horses lest they should ramble too far, and Eyre kept the first watch. The night was cold, the wind blowing a gale and driving the flying scud across the face of the moon. The horses wandered off in different directions in the scrub, giving the tired man much trouble to keep them together. About half-past ten he drove them near the camp intending shortly to call the overseer to relieve him.

Suddenly the dead stillness of the night and the wilderness was broken by the report of a gun. Eyre was not at first alarmed, thinking it was a signal of Baxter's to indicate the position of their camp. He called, but received no answer. Hastening in the direction of the shot, he was met by Wylie, the King George's Sound native, running towards him in great alarm crying out:—"Oh, massa, massa, come here!" and then losing speech from terror. Eyre was soon at the camp, and one glance was enough to see that his purpose must now be pursued grimly alone. Baxter, fatally wounded, was stretched upon the ground, bleeding and choking in his last agony. As Eyre raised his faithful companion in his arms he expired.

"At the dead hour of night, in the wildest and most inhospitable waste of Australia, with the fierce wind raging in unison with the scene of violence before me, I was left with a single native, whose fidelity I could not rely on, and who, for aught I knew might be in league with the other two, who, perhaps were even now lurking about to take my life, as they had done that of the overseer."

On examining the camp, Eyre found that the two boys had carried off both double-barrelled guns, all the baked bread and other stores, and a keg of water. All they had left behind was a rifle, with the barrel choked by a ball jammed in it, four gallons of water, forty pounds of flour, and a little tea and sugar.

When he had time to think the matter over calmly, Eyre judged, from the position of the body, that Baxter must have been aroused by the two natives plundering the camp, and that, getting up hastily to stop them, he was immediately shot. His first care was to put his rifle into serviceable condition, and then, when morning broke, he hastened to leave the ill-omened place. It was impossible to bury the body of his murdered companion; one unbroken sheet of rock covered the surface of the country for miles in every direction. Well might Eyre write, many years afterwards:—

"Though years have now passed away since the enactment of this tragedy, the dreadful horrors of that time and scene are recalled before me with frightful vividness, and make me shudder even now when I think of them. A lifetime was crowded into those few short hours, and death alone may blot out the impressions they produced."

The two murderers followed the white man and boy during the first day, evading all Eyre's attempts to bring them to close quarters, and calling to the remaining boy, Wylie, who refused to go to them. They disappeared the next morning, and must have died miserably of thirst and starvation.

Seven days passed without a drop of water for the horses, before they reached the end of the line of cliffs, and providentially came to a native well amid the sand dunes. From this point water was more frequently obtained, and what wretched horses they had left showed feeble symptoms of renewed life. At last, when their rations were completely exhausted, they sighted a ship at anchor in Thistle Cove. She proved to be the "Mississippi," commanded by Captain Rossitur, the whaler already referred to as the first foreign vessel to enter Port Lincoln; and once more Eyre had to give thanks for relief at a most critical moment.

For ten days, in the hospitable cabin of the French whaler, he forgot his sufferings, and regained some of his lost strength. Then, provided with fresh clothes and provisions, and with his horses freshly-shod, Eyre recommenced his weary pilgrimage, and, in July, 1841, arrived at his long-desired goal, King George's Sound.

In reflecting upon this painful march of Eyre's round the Great Bight, one feels an exceeding great pity that so much heroic suffering should have been spent on the execution of a purpose the fulfilment of which promised but little of economic value. The maritime surveys had fairly established the fact that no considerable creek or river found its way into the Southern Ocean, either in or about the Great Bight. Granted that the outflow of some of our large Australian rivers had been overlooked by the navigators, the local conditions were such as to render it virtually certain that any such omission was not made along this part of the south coast. Here there was to be found no fringe of low, mangrove-covered flats, studded with inlets and saltwater creeks, thus masking the entrance of a river. In some parts, a bold forefront of lofty precipitous cliffs, in others a clean-swept sandy shore, alone faced the ocean. Flinders, constantly on the alert as he was for anything resembling the formation of a river-mouth, would scarcely have been mistaken in his reading of such a coast-line. And the journey resulted in no knowledge of the interior, even a short distance back from the actual coast-line. The conjectures of a worn-out, starving man, picking his way painfully along the verge of the beach, were, in this respect, of little moment.

Eyre, however, won for himself well-deserved honour for courage and perseverance, in as exacting circumstances as ever beset a solitary explorer. The picture of the lonely man in his plundered camp bending over his murdered companion, separated from his fellow-men by countless miles of unwatered and untrodden waste, appeals resistlessly to our sympathies. But admiration of Eyre's good qualities has blinded many to his errors of judgment.

He was accorded a generous public welcome on his return to Adelaide, and was subsequently appointed Police Magistrate on the Murray, where his inland experience and knowledge of native character were of great service. When Sturt started on his memorable trip to the centre of Australia, Eyre accompanied his old friend some distance. But his activities were exercised in other fields than those of Australian exploration during his after life. He was Lieutenant-Governor of the Province of New Munster in New Zealand under Sir George Grey from 1848 to 1853, when that colony was divided into two provinces. He was afterwards Governor-General of Jamaica, where the active and energetic measures he took to crush the insurrection of 1865 incited a storm of opposition agaist him in certain quarters, and he played a leading part in the great constitutional cases of Philips v. Eyre, and The Queen v. Eyre. He died at Steeple Aston, in Oxfordshire, in 1906.

Chapter XII.

ATTEMPTS TO REACH THE CENTRE.

(i.)—Lake Torrens Pioneers and Horrocks.

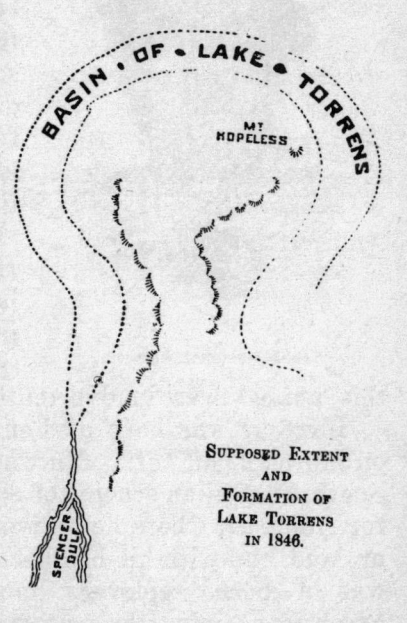

It will be remembered that Eyre, in 1840, reached, after much labour an elevation to the north-east, at the termination of the range which he had followed, and had named it Mount Hopeless. From the outlook from its summit he came to the conclusion that the lake was of the shape shown in the diagram, completely surrounding the northern portion of the new colony of South Australia. In fact, he formed a theory that the colony in far distant times had been an island, the low-lying flats to the east joining the plains west of the Darling. It was in 1843 that the Surveyor-General of South Australia, Captain Frome, undertook an expedition to determine the dimensions of this mysterious lake. He reached Mount Serle, and found the dry bed of a great lake to the eastward, as Eyre had described, but discovered that Eyre had made an error of thirty miles in longitude, placing it too far to the east. He got no further north. He thus confirmed the existence of a lake eastward of Lake Torrens (now Lake Frome), but achieved nothing to prove or disprove Eyre's theory of their continuity. Prior to this the pioneers had spread settlement both east and west of Eyre's track from Adelaide to the head of Spencer's Gulf. Amongst these early leaders of civilisation in the central state, are to be found the names of Hawker, Hughes, Campbell, Robinson, and Heywood. But unfortunately the details of their expeditions in search of grazing country have not been preserved.

John Ainsworth Horrocks is one of those, whose accidental death at the very outset of his career plunged his name into oblivion. Had he lived to climb to the summit of his ambition as an explorer, it would have been written large in Australian history. That he had some premonition of the conditions necessary to successful exploration to the west is shown by his having been the first to employ the camel as an aid to exploration. He took one with him on his last and fatal trip, and it is an example of fate's cruel irony that the presence of this animal was inadvertently the cause of his death.

John Ainsworth Horrocks.

Horrocks was born at Penwortham Hall, Lancashire, on March 22nd, 1818. He was very much taken with the South Australian scheme of colonisation, and left London for Adelaide, where he arrived in 1839. He at once took up land, and with his brother started sheep-farming. He was a born explorer, however, and made several excursions into the surrounding untraversed land, finding several geographical features, which still preserve the names he gave them. In 1846 he organised an expedition along more extended lines, intending to proceed far into the north-west and west. After having overlooked the ground, he would then prepare another party on a large scale to attempt the passage to the Swan River. He started in July, but in September occurred the disaster which cut him off in the flower of his promise. In his dying letter he describes how he saw a beautiful bird, which he was anxious to obtain:—

"My gun being loaded with slugs in one barrel and ball in the other, I stopped the camel to get at the shot belt, which I could not get without his lying down.

"Whilst Mr. Gill was unfastening it, I was screwing the ramrod into the wad over the slugs, standing close alongside of the camel. At this moment the camel gave a lurch to one side, and caught his pack in the cock of my gun, which discharged the barrel I was unloading, the contents of which first took off the middle fingers of my right hand between the second and third joints, and entered my left cheek by my lower jaw, knocking out a row of teeth from my upper jaw."

His sufferings were agonising, but he was easy between the fearful convulsions, and at the end of the third day after he had reached home, whither his companions had succeeded in conveying him, he died without a struggle.

(ii.)—Captain Sturt.

Charles Sturt, whose name is so closely bound up with the exploration of the Australian interior, had settled in the new colony which the South Australians loyally maintain he had created by directing attention to the outlet of the Murray. After a short re-survey of the river, from the point where Hume crossed it to the junction of the Murray and Murrumbidgee, which had been one of Mitchell's tasks, he re-entered civil life under the South Australian Government. He was now married, and settled on a small estate which he was farming, not far from Adelaide. In 1839 he became Surveyor-General, but in October of the same year he exchanged this office for that of Commissioner of Lands, which he held until 1843. In the following year he commenced his most arduous and best-known journey, a journey that has made the names of "Sturt's Stony Desert" and the "Depot

Glen" known all over the world, and that has, unhappily for Australia, done much to create the popular fallacy that the soil and climate of the interior are such as preclude comfortable settlement by whites. Sturt's graphic account is at times somewhat misleading, and the lapse of years has proved his denunciatory judgment of the fitness of the interior for human habitation to have been hasty. But if we examine the circumstances in which he received the impressions he has recorded, we must grant that he had considerable justification for his statements.

He was a broken and disappointed man, worn out by disease and frustrated hopes, and nearly blind. During six months of his long absence, he had been shut up in his weary depot prison, debarred from attempting the completion of his work, and compelled to watch his friend and companion die a lingering death from scurvy. And when the kindly rains released him, he was doomed to be repulsed by the ever-present desert wastes. No wonder that he despaired of the country, and viewed all its prospects through the heated, treacherous haze of the desert plains. Yet now, close to the ranges where Sturt spent the burning summer months of his detention, there has sprung up one of the inland townships of New South Wales, where men toil just as laboriously as in a more temperate zone.

But, though baffled and unable to win the goal he strove for, never did man better deserve success. The instructions that he received from the Home Office were, to reach the centre of the continent, to discover whether mountains or sea existed there, and, if the former, to note the flow and direction of the northern waters, but on no account to follow them down to the north coast. Sturt was instructed to proceed by Mt. Arden, a route already tried, condemned, and abandoned by Eyre; and he elected to proceed by way of the Darling. His plan was to follow that river up as far as the Williora, a small western tributary of the Darling, opposite the place whence Mitchell turned back in 1835, after his conflict with the natives, an episode which Sturt found that they bitterly

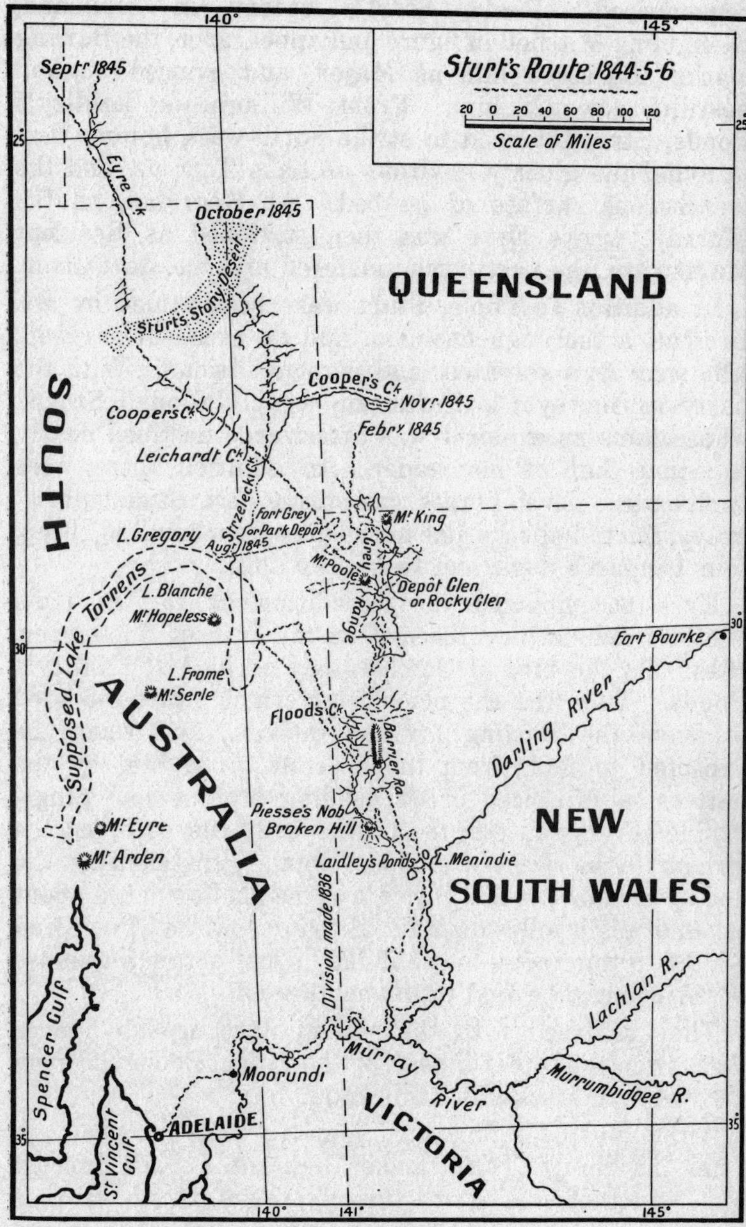
Sturt's Route 1844-5-6
20 0 20 40 60 80 100 120
Scale of Miles
140°
145°
141°
25°
30°
35°
Septr 1845
Eyre Ck
October 1845
Sturt's Stony Desert
QUEENSLAND
Cooper's Ck
Novr 1845
Febry 1845
Cooper's Ck
Leichardt Ck
Strzelecki Ck
Fort Grey or Park Depot
Augt 1845
Mt King
L. Gregory
Grey Range
Mt Poole
Depôt Glen or Rocky Glen
SOUTH
AUSTRALIA
Supposed Lake Torrens
L. Blanche
Mt Hopeless
Fort Bourke
L. Frome
Mt Serle
Darling River
Floods Ck
Barrier Ra.
Piesse's Nob
Broken Hill
NEW
SOUTH WALES
Mt Eyre
Mt Arden
Laidley's Ponds
L. Menindie
Division made 1836
Lachlan R.
Spencer Gulf
Murray River
Murrumbidgee R.
Moorundi
VICTORIA
St. Vincent Gulf
ADELAIDE

remembered. Poole, Sturt's second in command, resembling Mitchell in figure and appearance, the Darling blacks addressed him as Major, and evinced marked hostility towards him. From Williora, or Laidley's Ponds, Sturt intended to strike north-west, hoping thus to avoid the gloomy environs of Lake Torrens, and the treacherous surface of its bed. At Moorundi, on the Murray, where Eyre was then stationed as Resident Magistrate, the party was mustered and the start made.

In addition to Poole, Sturt was accompanied by Dr. Browne, a thorough bushman and an excellent surgeon, who went as a volunteer and personal friend. With the party as surveyor's draftsman, went McDouall Stuart, whose fame as explorer was afterwards destined nearly to equal that of his leader. In addition there were twelve men, eleven horses, one spring-cart, three bullock-drays, thirty bullocks, one horse-dray, two hundred sheep, four kangaroo dogs, and two sheep dogs.

Eyre accompanied the expedition as far as Lake Victoria, which they reached on the 10th of September, 1844. On the 11th of October they arrived at Laidley's Ponds. This was the place from which Sturt intended to leave the Darling for the interior, and where he expected to find, from the account given him by the natives, a fair-sized creek heading from a low range, visible at a distance to the north-west. But he found the stream to be a mere surface channel, distributing the flood water of the Darling into some shallow lakes about seven or eight miles distant. Sturt despatched Poole and Stuart to this range to see if they could obtain a glimpse of the country beyond to the north-west.

They returned with the rather startling intelligence that, from the top of a peak of the range, Poole had seen a large lake studded with islands.

Although in his published journal, written some time after his return, Sturt makes light of Poole's fancied lake, which of course was the effect of a mirage, at that time his ardent fancy, and the extreme likelihood of the existence of a lake in that locality, made him believe that he was on the eve of an important discovery. In a letter to Mr. Morphett of Adelaide, he wrote:—

"Poole has just returned from the range. I have not time to write over again. He says there are high ranges to the N. and N.W., and water, a sea, extending along the horizon from S.W. by S. and then E. of N., in which there are a number of lofty ranges and islands, as far as the eye can reach. What is all this? To-morrow we start for the ranges, and then for the waters, the strange waters, on which boat never swam and over which flag never floated. But both shall ere long. We have the heart of the interior laid open to us, and shall be off with a flowing sheet in a few days. Poole says that the sea was a deep blue, and that in the midst of it was a conical island of great height."

Poor Sturt! No boat was ever to float upon that visionary sea, nor flag to wave over those dream-born waters. To those who know the experiences that awaited the expedition, it is pathetic to read of the leader's soaring hopes, as delusive as the desert mirage itself.

The whole of the party now removed to a small shallow lakelet, the commencement of the Williora channel (Laidley's Ponds). After a short excursion to the distant ranges reported by Poole, Sturt, accompanied by Browne and two men, went ahead for the purpose of finding water of a sufficient permanency to remove the whole of the party to. At the small lake where they were then encamped, there was the ever-present likelihood of a conflict with the pugnacious natives of the Darling. He was successful in finding what he wanted, and on the 4th of November the main body of the expedition, finally leaving the Darling basin, removed to the new water depot.

The next day Sturt, with Browne and three men and the cart, started on another trip in search of water ahead. This was found in small quantities, but rain coming on, Sturt returned and sent Poole out again to search while the camp was being moved. On his return, Poole reported having seen some brackish lakes, and also having caught sight of Eyre's Mt. Serle. They were now well on the western slope of the Barrier Range, and, but for the providential discovery of a fine creek to the northward, which was called Flood's creek, after one of the party, they would have been unable to maintain their position. To Flood's creek the camp was removed, and Sturt congratulated himself on the steady and satisfactory progress he was making.

The party now left the Barrier Range, and followed a course to another range further north, staying for some time at a small lagoon while engaged in making an examination of the country ahead. On the 27th of January, 1845, they camped on a creek rising in a small range, and affording, at its head, a fine supply of permanent water. When upon its banks the explorers pitched their tents, they little thought that it would be the 17th of the following July before they would strike camp again. This was the Depot Glen, and an extract from Sturt's journal depicts the situation of the party:—

"It was not, however, until after we had run down every creek in the neighbourhood, and had traversed the country in every direction, that the truth flashed across my mind, and it became evident to me that we were locked up in the desolate and heated region into which we had penetrated, as effectually as if we had wintered at the Pole. It was long, indeed, ere I could bring myself to believe that so great a misfortune had overtaken us, but so it was. Providence had, in its all wise purposes, guided us to the only spot in that wide-spread desert where our wants could have been permanently supplied, but had there stayed our further progress into a region that almost appears to be forbidden ground."

This then was Sturt's prison—a small creek marked by a line of gum trees, issuing from a glen in a low range. By a kindly freak of nature, enough water had been confined in this glen to provide a permanent supply for the exploring party and their animals, during the long term of their detention.

Of Sturt's existence and occupation during this dreary period little can be said. He tried to find an avenue of

Photo by Rev. J. M. Curran.

Sturt's Depôt Glen. The Glen, eroded in vertical silurian slate, is less than a mile long. Poole rests by the creek where the gorge opens quite abruptly on to a vast cretaceous plain.

escape in every direction, until convinced of the futility of the attempt; sometimes encouraged and lured on by the shallow pools in some fragmentary creek, at others, seeing nothing before him but hopeless aridity. Now, too, he found himself attacked with what he then thought to be rheumatism, but which proved to be scurvy. Poole and Browne were afflicted in the same manner.

Sturt made one desperate attempt to the north during his imprisonment in the Depot Glen, and succeeded in reaching a point one mile beyond the 28th parallel, but further north he could not advance, nor did he find any inducement to risk the safety of his party.

There passed weeks of awesome monotony, relieved by one strange episode. From the apparently lifeless wilderness around them there strayed an old aboriginal into their camp. He was hungry and athirst, and in complete keeping with the gaunt waste from which he had emerged. The dogs attacked him when he approached, but he stood his ground and fought them valiantly until they were called off. His whole demeanour was calm and courageous, and he showed neither surprise nor timidity. He drank greedily when water was given to him, ate voraciously, and accepted every service rendered to him as a duty to be discharged by one fellow-being to another when cut off in the desert from his kin. He stopped at the camp for some time and recognised the boat, explaining that it was upside down, as of course it was, and pointing to the N.W. as the region where they would use it, thus raising Sturt's hopes once more. Whence he came they could not divine, nor could he explain to them. After a fortnight he departed, giving them to understand that he would return, but they never saw him again.

"With him," writes Sturt pathetically, "all our hopes vanished, for even the presence of this savage was soothing to us, and so long as he remained we indulged in anticipations for the future. From the time of his departure a gloomy silence pervaded the camp; we were indeed placed under the most trying circumstances; everything combined to depress our spirits and exhaust our patience. We had witnessed migration after migration of the feathered tribes, to that point to which we were so anxious to push our way. Flights of cockatoos, of parrots, of pigeons, and of bitterns; birds also whose notes had cheered us in the wilderness, all had taken the same road to a better and more hospitable region.''

And now the water began to sink with frightful rapidity, and all thought that surely the end must be near. Hoping against hope, Sturt laid his plans to start as soon as the drought broke up. He himself was to proceed north and west, whilst poor Poole, reduced to a frightful condition by scurvy, was to be sent carefully back to the Darling, as the only means of saving his life.

On the 12th and 13th of June the rain came, and the drought-beleaguered invaders of the desert were relieved. But Poole did not live to profit by the rain. Every arrangement was made for his comfort that their circumstances permitted, but on the first day's journey he died. His body was brought back and buried under the elevation which they called the Red Hill, and which is now known as Mount Poole, three and a-half miles from Depot Camp.

Sturt's way was now open. He again despatched the party selected to return to the Darling, whose departure had been interrupted by Poole's untimely death, and, with renewed hope, made his preparations for the long-denied north-west.

Having first removed the depot to a better grassed locality, he made a short trip to the west. On the 4th of August he found himself on the edge of an immense shallow, sandy basin, in which water was standing in detached sheets, "as blue as indigo, and as salt as brine." This he took to be a part of Lake Torrens. He returned to the new depot, called Fort Grey, which was sixty or seventy miles to the north-west of the Glen, and arranged matters for his final departure.

McDouall Stuart was left in charge of the depot. Dr. Browne accompanied the leader, and on the 14th of August a start was made. For some distance, owing to the pools of surface water left by the recent rain, they had no difficulty in keeping a straightforward course. The country they passed over consisted of large, level plains, intersected by sand-ridges; but they crossed numerous creeks with more or less water in all of them. To one of these creeks Sturt gave the name of Strzelecki. Finally they reached a well-grassed region which greatly cheered them with the prospect of success it held out. Suddenly they were confronted with a wall of sand; and for nearly twenty miles they toiled over successive ridges. Fortunately they found both water and grass, but the unexpected check to their brighter anticipations was depressing. Nor did a walk to the extremity of one of the ridges serve to raise their spirits.

Sturt saw before him what he describes as an immense plain, of a dark purple hue, with a horizon like that of the sea, boundless in the direction in which he wished to proceed. This was Sturt's "Stony Desert." That night they camped within its dreary confines, and during the next day crossed an earthy plain, with here and there a few bushes of polygonum growing beside some straggling channel in which they occasionally found a little muddy rain-water remaining. At night, when they camped just before dusk, they sighted some hills to the north, and, on examining them through the telescope, they discerned dark shadows on the faces, as if produced by cliffs. Next morning they made for these hills, in the hope of finding a change of country and feed for the horses, but they were disappointed. Sand ridges in repulsive array confronted them once more. "Even the animals," writes Sturt, "appeared to regard them with dismay."

Over plains and sand dunes, the former full of yawning cracks and holes, the party pushed on, subsisting on scanty pools of muddy water and fast-sinking native wells. On the 3rd of September, Flood, the stockman, who was riding in the lead, lifted his hat and waved it on high, calling to the others that a large creek was in sight.

Photo by Rev. J. M. Curran.

Poole's Grave and Monument, near Depot Glen, Tibbuburra, New South Wales.

When the main party came up, they feasted their eyes on a beautiful watercourse, its bed studded with pools of water and its banks clothed with grass. This creek Sturt named Eyre's Creek, and it was an important discovery in the drainage system of the region that he was then traversing.

Along this new-found watercourse, they were enabled to make easy stages for five days, when the course of the creek was lost; nor could any continuation be traced. The lagoons, too, that were found a short distance from the banks, proved to be intensely salt. Repeated efforts to continue his journey to other points of the compass only led Sturt amongst the terrible sandhills, their parallel rows separated by barren plains encrusted with salt. Sturt now came to the erroneous conclusion that he had reached the head of Eyre's Creek, and that further progress was effectually barred by a waterless tract of country. In fact, he was then within reach of a well-watered river, along which he could have travelled right up to the main dividing range of the northern coast. But Sturt was baffled in the most depressed area on the surface of the continent, where rivers and creeks lost their identity in the numberless channels into which they divided before reaching their final home in the thirsty shallows of the then unknown Lake Eyre. There was neither sign nor clue afforded him; his men were sick, and any further progress would jeopardise his retreat. There was nothing for it but to fall back once more; and, after a toilsome journey, they reached Fort Grey on the 2nd of October.

Sturt's last effort had been made to the west of north; he now made up his mind for a final effort due north. Before starting, however, he begged of Browne, who was still suffering, to retreat, while the way was yet open, to the Darling. This Browne resolutely refused to do; stating that it was his intention to share the fate of the expedition. The 9th of October saw Sturt again under way to the seemingly forbidden north, Stuart and two fresh men accompanying him. On the second day they reached Strzelecki Creek, and on the 13th they came on to the bank of a magnificent channel, with fine trees growing on its grassy banks, and abundance of water in the bed. This was the now well-known Cooper's Creek, which Sturt, on his late trip, had crossed unnoticed, as it was then dry and divided into several channels on their route. This was the most important discovery made in connection with the lake system, Cooper's Creek being one of the far-reaching affluents, its tributaries draining the inland slopes of the main dividing range.

Sturt, on making this unexpected discovery, was undecided whether to follow Cooper's Creek up to the eastward or persevere in his original intention of pushing to the north. A thunder-storm falling at the time made him adhere to his original determination, and defer the examination of the new river until his return.

Seven days after crossing Cooper's Creek, he had the negative satisfaction of seeing his gloomy forebodings fulfilled. Once more he gazed over the dreary waste of the "stony desert," unchanged and repellant as ever. They crossed it, but were again turned back by sandhill and salt plain, and forced to retrace their steps to Cooper's Creek. This creek Sturt followed up for many days, but found that it came from a more easterly direction than the route he desired to travel along; moreover, the one broad channel that they had commenced to follow became divided into several ana-branches, running through plains subject to inundation. This became so tiring to their now exhausted horses, who were wofully footsore, that he reluctantly turned back. He had found the creek peopled with well-nurtured natives, and the prospects of advancing were brighter than they had ever been; but both Sturt and his men were weak and ill, and the horses almost incapable of further effort. Moreover, he was not certain of his retreat.

As they went down Cooper's Creek on their way back, they found that the water was drying up so rapidly that grave fears were entertained lest Strzelecki's Creek, their main resource in getting back to Fort Grey, should be dry. Fortunately they were in time to find a little muddy fluid left, just enough to serve their needs. Here, though most anxious to get on, they were forced to camp the whole of one day, on account of an extremely fierce hot wind.

Sturt's vivid account of the day spent during the blast of that furnace-like sirocco has been oft quoted. But the reader should remember when reading it that the man who wrote it was in such a weakened condition that he had not sufficient energy left to withstand the hot wind, whilst the shade under which the party sought shelter was of the scantiest description.

They had still a distance of eighty-six miles to cover to get back to Fort Grey, with but little prospect of finding water on the way. After a long and weary ride they reached it, only to find the tents struck, the flag hauled down, and the Fort abandoned. The bad state of the water and the steady diminution of supply had forced Browne to fall back to Depot Glen. Riding day and night Sturt reached the old encampment, so exhausted that he could hardly stand after dismounting.

The problem of their final escape had now to be resolved. The water in Depot Creek was reduced so low that they feared there would be none left in Flood's Creek. If this failed, they were once more imprisoned. Browne, now much recovered, undertook the long ride of one hundred and eighteen miles which would decide the question. Preparations had been made for his journey by filling a bullock skin with water, and sending a dray with it as far as possible. On the eighth day he returned.

"Well, Browne," asked Sturt, who was helpless in his tent, "what news? Is it good or bad?" "There is still water in the creek," replied Browne, "but that is all I can say; what there is is as black as ink, and we must make haste, for in a week it will be gone.''

The boat that was to have floated over the inland sea was left to rot at Depot Glen. All the heaviest of the stores were abandoned, and the retreat of over two hundred miles commenced.

More bullock-skins were fashioned into water-bags, and with their aid and that of a scanty but kindly shower of rain, they crossed the dry stage to Flood's Creek in safety. Here they found the growth of the vegetation much advanced, and with care, and constant activity in searching ahead for water, they gradually increased the distance from the scene of their sufferings, and approached the Darling. Sturt had to be carried on one of the drays, and lifted on and off at each stopping-place. On the 21st of December, they arrived at the camp of the relief-party under Piesse, at Williorara, and Sturt's last expedition came to an end.

In taking leave of this explorer, we quote a short extract from his Journal to show the exalted character of the man whom Australians should ever regard with the greatest of pride:—

"Circumstances may yet arise to give a value to my recent labours, and my name may be remembered by after generations in Australia as the first who tried to penetrate to its centre. If I failed in that great object, I have one consolation in the retrospect of my past services. My path among savage tribes has been a bloodless one, not but that I have often been placed in situations of risk and danger, when I might have been justified in shedding blood, but I trust I have ever made allowance for human timidity, and respected the customs of the rudest people."

Sturt's health and eyesight had been greatly impaired by his last trip, but although he was for a time almost totally blind, he still managed to discharge the duties of Colonial Secretary. He was at last pensioned by the South Australian Government, and soon afterwards returned to England. He died at his residence at Cheltenham. Though the Home Office had treated him disgracefully during his life, and ignored his services, he lives for ever in the hearts of the Australians as the hero and chief figure of the exploration of their country. When he was on his death-bed, in 1869, the empty title of knighthood was conferred upon him. As he could not enjoy the tardy honour, his widow, who lived till 1887. was graciously allowed to wear the bauble.

Chapter XIII.

BABBAGE AND STUART.

(i.)—B. Herschel Babbage.

The unsolved problem of the extent and other details of that vast region of salt lakes and flat country then known under the generic name of Lake Torrens still greatly occupied the attention and excited the imaginations of the colonists of South Australia. And the accounts brought back by the different exploring parties were conflicting in the extreme. In 1851, two squatters, named Oakden and Hulkes, out run-hunting, pushed westward of Lake Torrens, and found suitable grazing country. They also discovered a lake of fresh water, and heard from the natives of other lakes to the north-west some fabulous legends of strange animals. Their horses giving in, Oakden and Hulkes returned, but although they applied for a squatting license for the country they had been over, it was not then settled or stocked. In 1856, Surveyor Babbage made some explorations in the field partly traversed by Eyre and Frome. He penetrated through the plains that were

B. Herschel Babbage.
Born 1815; died 1878.

supposed to occupy the central portion of the horseshoe formation at that time associated in the public opinion with Lake Torrens. More fortunate than his predecessors, he found permanent water in a gum-tree creek, and saw some fair-sized sheets of water, one of which he named Blanche Water, or Lake Blanche. Some further excursions led to the discovery of more fresh water and well-grassed pastoral country. The aboriginals, too, directed him to what they said was a crossing-place in that portion of Lake Torrens that had been sighted, in 1845, by Poole and Browne of Captain Sturt's party, when Poole thought he saw an inland sea. Their directions, however, proved unreliable, or Babbage failed to find the place, for he lost his horse in the attempt to cross the lake.

In 1857, another excursion to the westward of Lake Torrens was made by a Mr. Campbell, who discovered a creek of fresh water, which he called the Elizabeth. He also visited Lake Torrens, of which he reported in similar terms to those of previous explorers—that it was surrounded with barren country.

In April of the same year, a survey conducted by Deputy Surveyor-General Goyder, over the same country as that lately explored by Babbage, led to some absurd mistakes. A few miles north of Blanche Water he came to many surface springs surrounding a fine lagoon. To the north of them was an isolated hill, which he called Weathered Hill. From the summit of this hill he had a curious example of the effects of refraction in this region in a similar illusion to that which suggested Poole's inland sea. To the northward he saw a belt of gigantic gum-trees, and beyond them what appeared to be a sheet of water with elevated land on the far side. To the eastward, was another large lake. But all this was but the glamourie of the desert—on closer examination the gigantic gums dwindled down to stunted bushes, and the mountainous ground to broken clods of earth.

But the greatest surprise reserved for Goyder was at Lake Torrens, where he found the water quite fresh.

He described the Lake as stretching from fifteen to twenty miles to the north-west, with a water horizon, with an extensive bay forming to the southward; while to the north, a bluff headland and perpendicular cliffs were clearly to be discerned with the telescope. From the appearance of the flood-marks, Goyder came to the conclusion that there was little or no rise and fall in the lake, drawing the natural conclusion that its size was such as not to be influenced appreciably by flood waters, but that it absorbed them without showing any variation in its level.

Adelaide was overjoyed at the news. The threatening desert that hemmed in their fair province to the north was suddenly converted into a land of milk and honey. The Surveyor-General, Colonel Freeling, immediately started out, taking with him both a boat and an iron punt with which to float on these new waters. But there was a sudden fall to their hopes when a letter was received from him stating that the cliffs, the bay, and the headlands were all built up on the airy foundation of a mirage. The elves and sprites of this desolate region had been playing a hoax upon Goyder's party. But it is no wonder that Goyder had been so open to deception after unexpectedly finding fresh water in the lake that had been so long known as salter than the sea.

On reaching the lake, Freeling found the water still almost fresh; but one of Goyder's men who accompanied him, told him that it had already receded half-a-mile since the latter's visit. An attempt to float the punt was made, but after dragging it through mud and a few inches of water for a quarter of a mile, the men abandoned the attempt as hopeless. Freeling and some of the party then started to wade through the slush, but after proceeding three miles, and then sounding only six inches of water, they returned. Some of the more adventurous extended their muddy wade, but only met with a similar result. Lake Torrens was re-invested with its evil name, only somewhat shrunken in proportions.

In the same year, 1857, Stephen Hack started with a party from Streaky Bay to examine the Gawler Range of Eyre, and investigate the country west of Lake Torrens. He reached the Gawler Range and examined the country very carefully, finding numerous fresh-water springs, and large plains covered with both grass and saltbush. He also discovered a large salt lake, Lake Gairdner. Simultaneously with Hack's expedition, a party under Major Warburton was out in the same neighbourhood; in fact, Hack's party crossed Warburton's tracks on one or two occasions. Strange to say, the reports of the two were flatly contradictory. Warburton described the country as dry and arid; but Hack's account was distinctly favourable. Of the two men, however, it is most probable that Hack possessed the more experience and knowledge of country, and, moreover, Time, the great arbitrator, has endorsed his words.

The year 1857 saw much exploration done in South Australia. One party, consisting of Swinden, Campbell, Thompson, and Stock, at about seventy miles from the head of Spencer's Gulf, found good pastoral country and a permanent water-hole called by the natives "Pernatty." To the north they came upon Campbell's former discovery of the Elizabeth, but their provisions failing, they were forced to return.

A month afterwards Swinden started again from "Pernatty." North of the Gawler Range he found available pastoral country, which became known as Swinden's country. During this year also, Miller and Dutton explored the country at the back of Fowler's Bay. Forty miles to the north they saw treeless, grassy plains stretching far inland, but could find no permanent water. Warburton afterwards reported in depreciatory terms of this region; but Delisser and Hardwicke, who also visited it, stated that it would make first-class pastoral country if only surface water could be obtained. During the whole of Warburton's career, his judgment of the pastoral value of country seems to have been lamentably defective. He made no allowance for the varying nature of the seasons. A suggestion that he made to the South Australian Government to explore the interior, which had turned back such men as Sturt and Gregory, with the aid of the police, verges on the ludicrous.

In 1858, the South Australian Government voted a sum of money to fit out a party to continue the northern explorations. This party was put under the leadership of Babbage; but he was not given a free hand, being hampered with official instructions, and there being no allowance made for unforeseen exigencies. His instructions were to examine the country between Lakes Torrens and Gairdner, and to map the respective western and eastern shores of the two lakes, so as to remove for the future any doubt as to their actual formation and accurate position. This alone, apart from any extended exploration, meant a work of considerable time; but, unfortunately for the surveyor in charge, the general public was just then eager for fresh discoveries of available pastoral land, and was inclined to regard survey work as of secondary importance. It took several months to complete the survey work of the two lakes, and when Babbage returned to Port Augusta he found that Harris, the second in command of his depot camp, had started to return to Adelaide with many of the drays and horses. Babbage rode one hundred and sixty miles before he overtook him at Mount Remarkable, and there learned that the South Australian Government had changed its official mind with regard to the conduct of the expedition, and had decided that it should be conducted in future with pack-horses only.

It was A. C. Gregory's arrival in Adelaide with pack-horses from his last expedition down the Barcoo, that had led to this change of tactics. Charles Gregory, who had accompanied his brother, was now engaged by the Government to overtake Babbage and acquaint him with their intention, but when he reached Port Augusta, Gregory took it upon himself to order the drays home, Babbage being away surveying. Babbage overtook them and ordered them back; but pleading Government orders, they refused to return. Babbage wrote to the authorities pointing out the unfairness of their action, and, mustering up a small party, returned to continue his work with six months' provisions.

On this occasion, Babbage gave more time to discovery than he had done before. He went out beyond the boundaries of his survey, and pushed on to Chambers Creek, so called by Stuart, who discovered it while Babbage was busy at Lake Gairdner. Babbage traced Chambers Creek into Lake Eyre, and was thus the first discoverer of this lake, which he called Lake Gregory. He found a range which he called Hermit Range, but from its crest discerned no sign of Lake Torrens, thus settling a certain limit to its extension to the north. He made further explorations to the west of Lake Gregory, now Lake Eyre, and found some hot springs. Meanwhile, during the time he was making these researches, the Government had, in a very high-handed manner, appointed Warburton to supersede him. Warburton started out to find Babbage, taking Charles Gregory as his second. Failing to find him at the Elizabeth, he followed and overtook him at the newly-discovered Lake Gregory. Warburton made a few discoveries while seeking for Babbage, amongst them the Douglas, a creek which was afterwards of great assistance to Stuart, and the Davenport Range; and he also came upon some fair pastoral country.

Babbage's surveys and explorations had done much to clear up the mystery and confusion that had hitherto obscured the geography of the salt lake region. His discovery of Lake Eyre (Gregory) and of the complete isolation of Lake Torrens, reduced the component parts of that huge saline basin to some sort of method and order. In addition to these achievements, Surveyor Parry made some further discoveries both of fresh water and available pastoral country to the eastward of the Lake.

B. Herschel Babbage was the eldest son of the well-known inventor of the calculating machine. He had been educated as an engineer, and for a considerable time had followed his profession in Europe. He had been engaged on several main lines in England, and had worked in conjunction with the celebrated Brunel. He had also been commissioned by the Government of Piedmont to report on a line across the Alps by way of Mount Cenis. He had remained in Italy some years until his work was interrupted by the revolution. He had returned to England, and had subsequently come to South Australia in 1851, in the ship "Hydaspes." He died at his residence, in 1878, at St. Mary's, South Road, where he had a vineyard.

(ii.)—JOHN MACDOUALL STUART.

John MacDouall Stuart, the great explorer of the centre of Australia, arrived in South Australia in 1839. His first experience of Australian exploration was sufficiently trying, gained as it was when he was acting as a draughtsman with Captain Sturt on his last arduous expedition. But it had kindled in him a high ardour for discovery, and fostered a stubborn resolution to carry through whatever he undertook.

He commenced his early explorations when in a position to do so independently, to the north-west of Swinden's country, in search of some locality called by the natives "Wingillpin." Not finding it, he came to the strange conclusion that Wingillpin and Cooper's Creek were one and the same, although he was now on a different watershed. He also, at that period, seems to have entertained somewhat extensive notions of the course of Cooper's Creek, as in one part of his Journal he remarks:—

"My only hope of cutting Cooper's Creek is on the other side of the range. The plain we crossed to-day resembles those of the Cooper, also the grasses. If it is not there, it must run to the north-west, and form the Glenelg of Captain Grey."

Now, although we know that Grey held rather extravagant notions of the importance of the Glenelg, even he would not have thought it possible for the Glenelg to be the outlet of such a mighty river as Cooper's Creek would have become by the time it reached the north-west coast.

Stuart's horses were now too footsore to proceed over the stony country he found himself then in, and he had no spare shoes with him. Failing therefore to find the promised land of Wingillpin, although he had passed over much good and well-watered country, he turned to the south-west, and made some explorations in the neighbourhood of Lake Gairdner. Before this, however, he had found and named Chambers Creek. From Lake Gairdner, he steered for Fowler's Bay, and his description of some of the country he passed is anything but inviting. From a spur of the high peak that he named Mount Fincke, he saw :—

"A prospect gloomy in the extreme: I could see a long distance, but nothing met the eye save a dense scrub, as black and dismal as night."

From this point the party passed into a sandy, spinifex desert, which Stuart says was worse than Sturt's; there had been a little salt-bush there, but here there was nothing but spinifex to be found, and the barren ground provided no food of any kind for the horses.

The state of affairs was becoming desperate with the little band, as their provisions were nearly finished; and though the leader was tempted to persist in the search for good pastoral country, he was at last forced to abandon the search and beat a hasty retreat. Dense scrub and the same "dreary dismal desert," as he calls it in his Journal, surrounded them day after day. Tired out and half-starved they reached the coast, and had but two meals left to carry them to Streaky Bay, where they

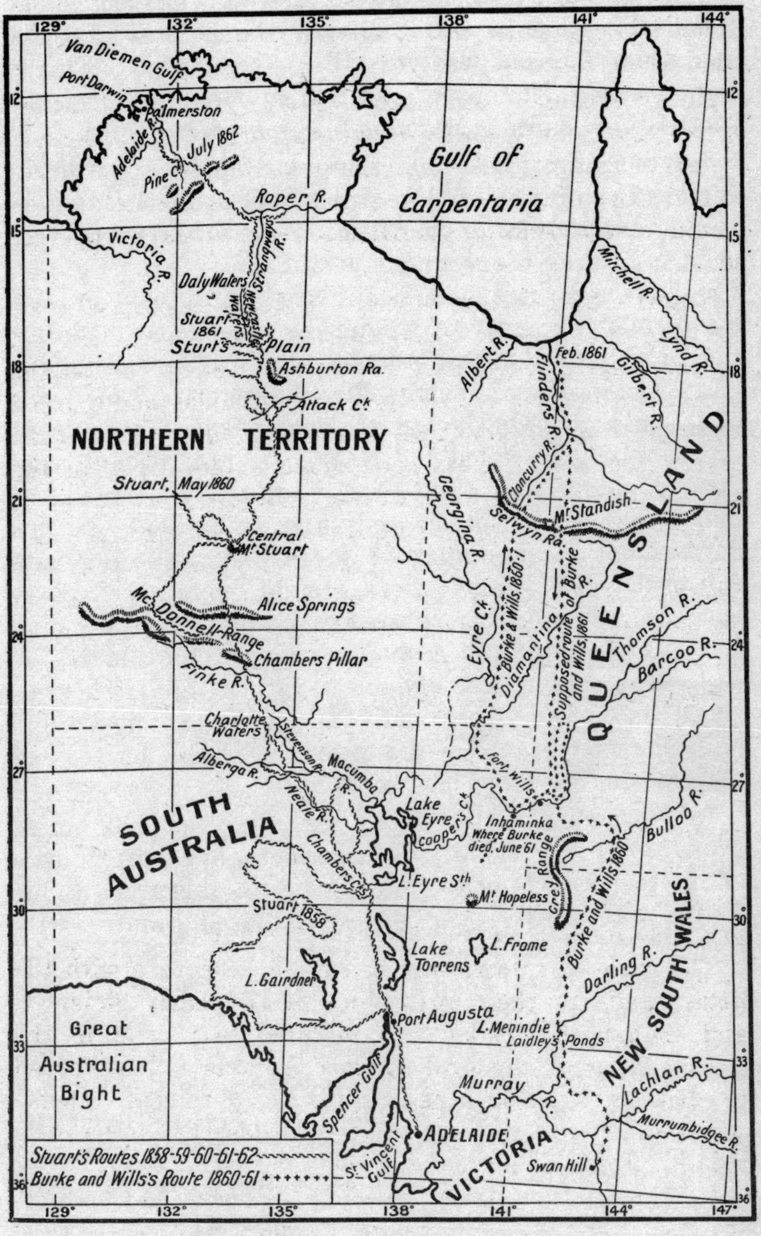
Van Diemen Gulf
Port Darwin
Palmerston
Adelaide R.
July 1862
Pine Ck.
Roper R.
Gulf of Carpentaria
Victoria R.
Daly Waters
Strangways R.
Newcastle Waters
Stuart 1861
Sturt's Plain
Ashburton Ra.
Attack Ck.
NORTHERN TERRITORY
Stuart, May 1860
Central Mt. Stuart
Alice Springs
McDonnell Range
Chambers Pillar
Finke R.
Charlotte Waters
Stevenson R.
Alberga R.
Macumba R.
Neale R.
SOUTH AUSTRALIA
Lake Eyre
Cooper's Ck.
Chambers Ck.
L. Eyre Sth
Stuart 1858
Mt. Hopeless
L. Frome
Lake Torrens
L. Gairdner
Port Augusta
Great Australian Bight
Spencer Gulf
St. Vincent Gulf
ADELAIDE
VICTORIA
Swan Hill
Murray R.
Murrumbidgee R.
Lachlan R.
NEW SOUTH WALES
L. Menindie
Laidley's Ponds
Darling R.
Burke and Wills 1860
Grey Range
Inhaminka
Where Burke died, June '61
Fort Wills
Bulloo R.
Barcoo R.
Thomson R.
QUEENSLAND
Supposed route of Burke and Wills, 1861
Diamantina R.
Burke & Wills 1860-1
Eyre Ck.
Georgina R.
Selwyn Ra.
Mt. Standish
Cloncurry R.
Flinders R.
Feb. 1861
Albert R.
Gilbert R.
Lynd R.
Mitchell R.
Stuart's Routes 1858-59-60-61-62
Burke and Wills's Route 1860-61

found relief at Gibson's station. Here the sudden change
from starvation to a full diet invalided most of them,
and Stuart himself was very ill for some days. Finally
they reached Thompson's station at Mount Arden, and
there Stuart's first expedition terminated.

But this severe test only whetted Stuart's appetite for
further exploration, and in April, 1859, he made another
start. After crossing over some of the already-traversed
country, Hergott, one of his companions, found the now
well-known springs that bear his name. Stuart crossed
his former discovery of Chambers Creek, and made for
the Davenport Range, discovered by Warburton, finding
many of the mound springs that characterize some parts
of the interior. On the 6th of June he discovered a large
creek, which he called the Neale. It ran through very
good country, and Stuart followed it down, hoping to
find it increase in volume and value as he went. In this
he was not disappointed, as large plains covered with
salt-bush and grass were found, and the party
encountered several more springs. After satisfying himself of the extent and economic value of the country he
had found, Stuart was obliged to return; for his horses'
shoes had again worn out, and he had a lively and
painful remembrance of the misery which his horses had
suffered before from lack of them.

In November of the same year, he made a third
expedition in the vicinity of Lake Eyre, but there is little
of interest attaching to the Journal of this trip, as his
course was mostly over closely explored country. He
reached the Neale again, and instituted a survey of the
promising pastoral country he had traversed during his
last trip, approaching at times to within sight of what he
calls in his Journal Lake Torrens, but which in reality
was what is now known as Lake Eyre. All these minor
expeditions of Stuart's may be looked upon as
preparatory to his great struggle to find an available
passage through the unknown fastnesses of the centre of
the continent.

It was in 1860 that Stuart made the first of his daring and stubborn attempts to cross Australia from south to north. The South Australian Government had offered a standing reward of £2,000 for the man who should first succeed in this undertaking.

Stuart's party on his first trip was but a very small one: three men in all, with but thirteen horses. It reads liliputian compared with the princely cavalcade that later on set out with Burke to travel over comparatively well-known country, involving only a short excursion through a land without natural difficulties or obstacles; and yet it actually achieved the greatest part of the task set it.

Stuart started from Chambers Creek, but for part of the journey he was of course travelling over country that was fairly well-known by that time. After passing the Neale, he entered untrodden country, which proved to be good available pastoral land. Numerous well-watered creeks were passed, which were named respectively the Frew, the Fincke, and the Stevenson, and on the 6th of April they reached a hill of a remarkable shape, which had for some time attracted and excited their attention and curiosity. They found it to be a column of sandstone, on the apex of a hill. The hill was but a low one of a few hundred feet in height, but the sandstone column that surmounted it was one hundred and fifty feet in height and twenty feet in width. This striking object was named by Stuart Chambers Pillar, to commemorate a friend who had assisted him greatly in his explorations. It stood amongst other elevations of fantastic shapes and grotesque formations, resembling ruined forts and castles. On the 9th of April they sighted two remarkable bluffs, and on the 12th reached the range of which the bluffs formed the centre. The eastern bluff was called Brinkley Bluff and the western Hanson Bluff; the range, which is now well-known as a leading geographical feature of Australia, and on which the most elevated peaks in the interior have since been found, Stuart named the MacDonnell Range, after the then Governor of South

Australia. The little band crossed the range, which was rough but had good grass on its slopes. There was, however, a scarcity of water; for they were now approaching the tropical line, and on reaching the northern slope of the range found themselves amongst spinifex and scrub, and obliged to undergo two nights without water for the horses. At a high peak, which was named Mount Freeling, they found a small supply; and as it was now evident that there was dry country ahead, a more careful search was made before pushing any further forward, in order to ensure certain means of retreat. Fortunately they found, amongst some ledges of rock, a large natural reservoir, which promised to be permanent, and capable of supplying their wants on their homeward way.

On the 22nd of April, Stuart camped in the centre of Australia, on the spot which his former leader, Sturt, had vainly undergone so much suffering to reach; and his feeling of elation must have been tempered with regret that his old leader was not then with him to share this success. About two miles and a-half to the North-North-East there was a tolerably high hill which he called in reality Central Mount "Sturt." It is now, however, erroneously called "Stuart," owing to the publishers of his diary having misread his manuscript.

Having, in company with his tried companion Kekwick, climbed the mount, he erected a cairn of stones at the top and hoisted the Union Jack. They then recommenced their northern journey. That night they camped without finding water, but the next morning were lucky enough to get a permanent supply. Then ensued much delay, caused by fruitless attempts to strike either to the eastward or the westward. Stuart tried on several occasions to reach the head of the Victoria River, but failed, and sacrificed some horses. On a creek he called the Phillips, some natives were encountered who, according to Stuart, made and answered a masonic sign.

To the north of this spot, the explorers came to a large gum-tree creek, with very fair-sized sheets of water in it.

As they followed down, they passed an encampment of natives, but kept steadily on their course without interfering with them. Not finding any water lower down the creek, the party had to return, and when close to the creek at the point where they had crossed that morning, they were suddenly surrounded by a mob of armed and painted savages, who had emerged unexpectedly from concealment in a clump of scrub. To all attempts at peaceful parley they returned showers of boomerangs and clubs, until the whites were compelled in self-defence to fire on them. Even then they were not deterred from following the party, even up to the camp of the night before. This incident caused Stuart to hesitate. His party was so small that the loss or even disablement of one man would have crippled the expedition; and they had already lost a good many horses. He therefore wisely decided to fall back, as they had penetrated far enough to prove that the passage of the continent could be effected with a few more men. It was on the 27th of June that he began his homeward march, and on the 26th of August he reached Brodie's camp at Hamilton Springs, with the strength of all much reduced, and Stuart himself suffering from scurvy.

After the result of Stuart's journey had been reported in Adelaide, and it was seen how inadequate means only had led to his defeat, the Government voted £2,500 to equip a better-organized party; of this he was to take command.

Stuart judged it best to keep his old track by way of the Fincke and the Hugh. On the 12th of April they arrived at the Bonney, and finding it running strong, with abundance of good feed on the banks, they were betrayed into following it down; but it soon spread abroad and was lost in a large plain. Leaving the Bonney, they adhered to the old route, and reached Tennant's Creek on the 21st of April, and four days afterwards they were on the scene of the attack that had been made on them at Attack Creek. But although the tracks of the natives were numerous, the explorers were, at this time, permitted to pass on in peace. Keeping at the foot of the low range, which there has an approximate northerly and southerly direction, Stuart crossed many creeks which promised long courses where they formed in the range, but which were all alike lost when they reached the level country. On the 4th of May they attained to the northern termination of this range, which he called the Ashburton Range. Here he made several attempts to the north and north-west, but could discover neither water nor watercourses in those directions; nothing indeed but plains, beautifully grassed, but heavy to ride over and yielding under the horses' feet. Beyond these plains, the country changed for the worse, and became sandy and scrubby. On the 16th of May he encountered a new description of scrub that grew in a very obstructive manner, and is now known as Stuart's "Desert Hedgewood."

On the 23rd he found a magnificent sheet of permanent water which he called the Newcastle Waters, and at first he judged that a clear way north was now assured. But he was deluded, for beyond these waters he could not advance his party a mile; north, north-east, and north-west, there was the one outlook—endless grassy plains, terminating in dense scrubby forest country. He had to give up all hope for the present, and return to Adelaide.

Such however was the confidence of the authorities in him, and such his own energy, that in less than a month after his arrival in Adelaide he was on his way to Chambers Creek to make preparations for a fresh departure. His last two journeys had proved the existence of a long line of good country, fairly well-watered; and although beyond it he had not been able to gain a footing, still there was no knowing what a fresh endeavour would bring to light.

He had brought his party back in safety, with the loss of only a few horses, and had actually reached in point of position as low a latitude as the Victorian explorers had done, and that with a more difficult country to travel through, without camels, and with an inferior equipment in all other respects.

It is not necessary again to follow Stuart's horse-tracks over the northern way he was now pursuing for the third time. On the 14th of April, 1862, we find him encamped at the northern end of Newcastle Waters, once more about to force a passage through the forest of waterless scrub to the north. On the second day he was partly successful, finding an isolated waterhole, surrounded by conglomerate rocks. This he called Frew's Pond; and it is now a well-known camping-place for travellers on the overland telegraph line.

Past this spot he was not able to make any progress. Twice he made strenuous but vain efforts to reach some tributary of the Victoria River. He then spent many days riding through dense mulga and hedgewood scrub. At length, after much hope deferred, finding a few scanty waterholes that did not serve the purpose he had in view, he succeeded in striking the head of a chain of ponds running in a northerly direction. These being followed down, led him to the head of the creek now called Daly Waters Creek, and finally to the large waterhole on which the present telegraph station bearing the name of Daly Waters, stands. The creek was then lost in a swamp, and Stuart was unable to find the channel where it reformed, which has since been named the Birdum. Missing this water-guide, Stuart worked his way to the eastward, to a creek he named the Strangways, which led him down to the Roper River, a river which he had never striven to reach, his sole aim being the Victoria. He crossed the Roper, and followed up a northern tributary, which he named after his constant friend John Chambers.

Scarcity of water was now a thing of the past, but his stock of spare horseshoes had to be most jealously guarded, for his horses were beginning to fall lame, the country he was on was very stony, and he was far removed from Adelaide. From the Chambers he came to the lower course of a creek called by Leichhardt Flying-Fox Creek, re-named by Stuart the Katherine, the name it now bears. Thence he struck across the stony tableland and descended on the head waters of a river which he christened the Adelaide, and on following this river down he found himself in rich tropical scenery, which told him that at last he was approaching the sea-shore.

On the 24th of July he turned a little to the north-east, intending to strike the sea-beach and travel along it to the mouth of the Adelaide. He told only two of the party of the eventful moment awaiting them. As they rode on, Thring, who was riding ahead, suddenly called out, "The Sea," which so took the majority by surprise that they were some time before they understood what was meant, and then three hearty cheers were given.

At this, his first point of contact with the ocean, Stuart dipped his feet and hands in the sea, as at last he gazed across the water he had so perseveringly striven for years to reach.

He attempted to get to the mouth of the Adelaide River along the beach, but found it too boggy for the horses. Wishing to husband the forces at his command, Stuart wisely resolved to push no further; he had a space cleared where they were, and a tall sapling stripped of its boughs to serve as a flagstaff. On this he hoisted the Union Jack which he had carried with him. A record of their arrival, contained in an air-tight case, was then buried at the foot of the impromptu staff, and Stuart cut his initials on the largest tree he could find. The tree has since been found and recognised, but the buried memorial has not been discovered. More fortunate than the ill-fated Burke, Stuart surveyed the open sea from his point of contact with the ocean, instead of having to be content with some mangrove trees and salt water.

McDouall Stuart, whose last expedition we have thus followed out to its successful end, is rightly considered the man to whom the credit for first crossing the continent is due. His victory was all his own; he had followed in no other person's footsteps; he had crossed the true centre, and he had made the coast at a point much further to the north than that reached by

Burke and Wills, their journey having been considerably shortened by its northern end being placed on the southern shore of the great gulf that bites so deeply into north Australia. Along Stuart's track there is now erected the Overland Telegraph Line, an enduring monument to his indomitable perseverance.

Stuart's health was fast failing, and his horses were sadly reduced in strength. He therefore started back the day after the consummation of his dearest ambition. On his way south, after leaving Newcastle Waters, he found the water in many of the short creeks heading from the Ashburton Range to be rapidly diminishing; in some there was none left, in others it was fast drying. The horses commenced to give in rapidly one after the other, and more were lost on successive dry stages. Stuart himself thought that he would never live to see the settled districts. Scurvy had brought him down to a lamentable state, and after all his hard-won success, it seemed as though he would not profit by it. His right hand had become useless to him, and his eyes lost power of sight after sunset. He could not undergo the pain of riding, and a stretcher had to be slung between two horses to carry him on. With painful slowness they crept along until they reached Mount Margaret, the first station. Here the leader, reduced to a mere skeleton, was furnished with a little relief; and after resting and gaining a little strength, he rode on to Adelaide.

This was Stuart's last expedition; for he never recovered his health nor former eyesight. He was rewarded by the Government of the colony which he had served so well, and was awarded the gold medal of the Royal Geographical Society. He went to reside in England, where he died in the year 1869, on the 16th of July.

Chapter XIV.

BURKE AND WILLS.

We have now to deal with an exploring expedition of greater notoriety than that of any similar enterprise in the annals of Australia, though its results in the way of actual exploration in the true meaning of the term were quite insignificant. The expedition could not reasonably hope to reveal any new geographical conditions; for the nature of the country to be traversed was fairly well-known: there was no such expanse of unknown territory along the suggested course of travel as to justify the anticipation of any discovery of magnitude. Both Kennedy and Gregory had followed much the same line of route when tracing the course of the Barcoo and Cooper's Creek, a short distance to the eastward. The only apparent motive for the expedition seems to have been not particularly creditable, the desire to outdo Stuart, who after nearly accomplishing the task might well have been allowed the honour of completing it. But Time is after all the great arbitrator: Stuart re-entered Adelaide successful, on the same day that the bodies of Burke and Wills arrived for shipment to Melbourne.

Robert O'Hara Burke was born in the county of Galway, in Ireland, in 1821. He was the second son of John Hardiman Burke, of St. Clerans, and was educated in Belgium. In 1840 he entered the Austrian army, in which he rose to the rank of Captain. In 1848 he joined

Robert O'Hara Burke.
From a photograph in the possession of E. J. Welch, of the Howitt Relief Expedition.

the Royal Irish Constabulary, but five years later emigrated to Tasmania. Thence he went to Victoria, where he entered the local police force, and became an Inspector. Such was his position when he was offered the command of the expedition which ended in his death.

William John Wills was born at Totnes, in Devonshire. He was the son of a medical man, and after his arrival in Victoria, in 1852, he led for a time a bush life on the Edwards River. He was later employed as a surveyor in Melbourne, and then became assistant to Professor Neumayer at the Melbourne Observatory, a post he quitted in order to act as assistant-surveyor on the ill-starred journey.

William John Wills.
From a photo in possession of E. J. Welch, of the Howitt Relief Expedition.

Sentiment, and an hysterical sentiment at that, seems to have dominated this expedition throughout. There was no urgent necessity for Victoria to equip and send forth an exploring expedition. Her rich and compact little province was known from end to end, and she had no surplus territory in which to open up fresh fields of pastoral occupation for her sons. But her people became possessed with the exploring spirit, and the planning and execution of the scheme was a signal indication of national patriotism. And if sense and not sentiment had marked the counsel, the results might have conferred rich benefit upon Australia.

The necessary funds were made up as follows:— £6,000 voted by Government; £1,000 presented by Mr. Ambrose Kyte; and the balance of the first expenditure of £12,000 made up by public subscription. But the final cost of the expedition and of the relief parties amounted to £57,000. And the exploratory work done by the different relief parties far and away exceeded in geographical results the small amount effected by the original expedition.

A committee of management was appointed, and to his interest with this committee Burke owed his elevation to the position of leader. He seems to have been supported by that sort of general testimony which fits a man to apply for nearly any position; but of special aptitude and training for the work to be done he had none. He was frank, openhearted, impetuous, and endowed with all those qualities which made him a great favourite with women; moreover, his service in the Austrian army had given people an exaggerated notion of his ability to command and organize. It would appear on the whole that his appointment was due solely to the influence he wielded, and to his personal popularity.

Wills appears to have been a man gifted with many of the qualities essential for efficient discharge of the duties and responsibilities appertaining to the post he held; but his amiable disposition allowed him to be influenced too readily in council by the rash and foolish judgment of his impetuous superior. If, for instance, he had persisted in combating Burke's incomprehensible plan of leaving the depot for Mount Hopeless, the last fatality would never have occurred.

When the expedition left Melbourne, it was amid the shouts and hurrahs of acclaiming thousands, who probably had not the faintest idea of the easy task that the explorers with their imposing retinue and outfit had before them. In fact, with all the resources at Burke's command, a favourable season and good open country, the excursion would have been a mere picnic to most men of experience. A number of camels had been specially imported from India at a cost of £5,500. G. J. Landells came to the country in charge of them, and had been appointed second in command. Long before they left the settled districts, Burke quarrelled with him, whereupon he resigned and returned to Melbourne. There he openly declared that under Burke's control the expedition would assuredly meet with disaster. Wills was then appointed second by Burke, and Wright, who was supposed to be acquainted with the locality which they were approaching, was engaged as third, another most unfortunate selection. Besides those already mentioned, there were Dr. Hermann Beckler, medical officer and botanist, and Dr. Ludwig Beckler, artist, naturalist, and geologist, ten white assistants, and three camel-drivers.

The expedition in full reached Menindie on the Darling, where Wright joined them. On the 19th of October, 1860, Burke, Wills, six men, five horses and sixteen camels, left Menindie for Cooper's Creek. Wright went with them two hundred miles to indicate the best route, and then returned to take charge of the main body waiting at Menindie. On the 11th of November, Burke with the advance party reached Cooper's Creek, where they camped and awaited the arrival of Wright with the rest. Grass and water were both plentiful, and the journey had hitherto proved no more arduous than an ordinary overlanding trip.

The long delay and inaction worked sadly upon Burke's active and impatient temperament, and he suddenly announced his intention to subdivide his party and, with three men, to start across the belt of unknown country— a distance of five hundred miles at the furthest—that separated him from Gregory's track round the Gulf. Although his lavish outfit had been purchased specially to explore this comparatively small extent of land, he thus deliberately left it behind him during the most critical part of the journey. He had with him no means of following up any discoveries he might make, and his botanist and naturalist and geologist were also left behind. He killed time for a little while by making short excursions northward, and then, on the 16th of December, impatient of further delay, he started with Wills and two men for Carpentaria. The others were left, with verbal instructions, to wait three months for him. Thus, dispersed and neglected, he left

the costly equipment containing within itself all the
elements of successful geographical research. Certainly
this was not the plan that had been anticipated by the
promoters and organisers. We have now, at this stage,
the spectacle of the main body loitering on the outskirts
of the settled districts, four men killing time on the banks
of Cooper's Creek, and the leader and three others
scampering across the continent, all four of them utterly
inexperienced in bushcraft.

As might have been expected the results of the journey
are most barren: Wills's diary is sadly uninteresting, and
Burke made only a few scanty notes, at the end of which
he writes:—"28th March. At the conclusion of report
it would be as well to say that we reached the sea, but we
could not obtain a view of the open ocean, although we
made every endeavour to do so."

Shortly condensing Wills's diary, we gather the
following account of their route. The first point they
intended to reach was Eyre's Creek, but before arriving
at it, they discovered a fine watercourse coming from
the north, which took them a long distance in
the direction they desired to follow. This watercourse,
which McKinlay afterwards called the Mueller, began in
time to lead their steps too much to the eastward, in which
direction lay its source. They therefore quitted it and
kept due north, following a tributary well-supplied with
both grass and water. This tributary led them well on
to the northern dividing range, which they crossed
without difficulty, coming down on to the head of the
Cloncurry River. By tracing that river down they
reached the Flinders River, which they followed down to
the mangroves and salt water. They were, however,
considerably out in their longitude, for they thought
that they were on the Albert, over one hundred miles to
the westward.

Having sighted salt water, if not the open sea, they
commenced the retreat. Gray and King were the two
men who were with Burke and Wills; and for equipment
they had started with six camels, one horse, and three

Scenes on Cooper's Creek. (After Howitt.)

months' provisions. Short rations and fatiguing marches now began to tell, and during the struggle back to the Depot, there seems to have been an absence of that kindly spirit of comradeship that has so often distinguished other exploring expeditions fallen on evil days.

Gray became ill, and took some extra flour to make a little gruel with. For this infringement of rules, Burke personally chastised him. A few days afterwards, Wills wrote in his diary that they had to halt and send back for Gray, who was "gammoning" that he could not walk. Nine days afterwards the unfortunate man died, an act which is not often successfully "gammoned."

But to bring the miserable story to an end, at last on the evening of the 21st of April, 1861, two months after they had reached the Gulf, they re-entered the depot camp at Cooper's Creek, where four men had been instructed to await their return, only to find it deserted and lifeless. Keenly disappointed, for though they knew they were behind the appointed time, they had still hoped that some one would have waited for them, they searched the locality for some sign or message from their friends, and on a tree saw the word "DIG" carved. Beneath this message of hope they were soon busy digging, and before long they unearthed a welcome store of provisions and a letter, which ran:—

"Depot, Cooper's Creek, April 21, 1861.—The depot party of V.E.E.* leaves this camp to-day to return to the Darling. I intend to go S.E. from Camp 60 to get on our old track at Bulloo. Two of my companions and myself are quite well; the third—Patton—has been unable to walk for the last eighteen days, as his leg has been severely hurt when thrown by one of the horses. No person has been up here from the Darling. We have six camels and twelve horses in good working condition.

WILLIAM BRAHE."

Unfortunately, this was so worded that when Burke found it the same night, it gave him the impression that the depot party were all, with one exception, fairly well;

* Victorian Exploration Expedition.

and that, with fresh animals just off a long rest they would travel long stages on their homeward march. As a matter of fact, on the evening of the day that Burke returned, they were camped but fourteen miles away. But this was only the first of a series of singular and fatal oversights—that almost seemed pre-ordained by mocking Fate.

Burke consulted his companions as to the feasibility of their overtaking Brahe, and they both agreed that, in their tired and enfeebled condition, it was hopeless to attempt it. Burke proposed that instead of returning up the creek along the old route to Menindie, they should follow the creek down to Mount Hopeless in South Australia, following the route taken by A. C. Gregory.* Wills objected to this, and so did King, but ultimately both gave in, thereby signing their death warrant; for if they had remained quietly at the depot, they would have been rescued.

After resting for five days, and finding their strength much restored by the food, they started for Mount Hopeless, ill-omened name. Before they left, Burke placed in the cache a paper, stating that they had returned, and then carefully restored the ground to its former condition. The common and natural thought to mark a tree or to make some other unmistakable sign of their return, does not seem to have occurred to either of the leaders. It will be seen further on how this scarcely credible omission was a main factor in deciding their fate.

As they progressed slowly down the creek, one of the two camels became bogged, and had to be shot where it lay. The wanderers cut off what meat there was on the body, and stayed two or three days to dry it in the sun. The one camel had now to carry what they had, except the bundles that the men bore, each some twenty-five pounds in weight. They made but little progress; the creek split up into many channels that ran out into earthy plains; and at last, when their one beast of burden gave in, they had to acknowledge defeat, and commenced to return. After shooting the wretched camel and drying

* See page 253.

his flesh, the men tried to live like the blacks, on fish and
"nardoo," the seeds of a small plant of which the natives
make flour. But the struggle for existence was very hard;
they were not expert hunters, and the natives, who were
at first friendly and shared their food with them, soon
out-grew the novelty of their presence, began to find them
an encumbrance, and constantly shifted camp to avoid
the burden of their support.

On the 27th of May, Wills went forward alone to visit
the depot and deposit there the journals and a note
stating their condition. He reached there on the 30th and
wrote in his diary that "No traces of anyone, except
blacks have been here since we left."

But while they were absent down the creek, Brahe and
Wright had visited the place, and finding no sign of their
return, and the cache apparently untouched, had ridden
away concluding that they had not yet come back. This
was the note that Wills left:—

"May 30th, 1861. We have been unable to leave the
creek. Both camels are dead. Burke and King are down
on the lower part of the creek. I am about to return to
them, when we shall probably all come up here. We are
trying to live the best way we can, like the blacks, but we
find it hard work. Our clothes are going fast to pieces;
send provisions and clothes as soon as possible.

"The depot party having left contrary to instructions
has put us into this fix. I have deposited some of my
journals here for fear of accidents."

WILLIAM J. WILLS.

Having done this, and once more carefully concealed
all traces of the cache having been disturbed, Wills
rejoined his companions in misfortune. Some friendly
natives fed him on his way back to them.

During the intercourse that of necessity they had with
the natives along Cooper's Creek, they had noticed the
extensive use made by them of the seeds of the nardoo
plant; but for a long time they had been unable to find
this plant, nor would the blacks show it to them. At
last King accidentally found it, and by its aid they

managed to prolong their lives. But the seeds had to be gathered, cleaned, pounded and cooked; and in comparison with all this labour the nourishment afforded by the cakes was very slight. An occasional crow or hawk was shot, and a little fish now and then begged from the natives. As they were sinking rapidly, it was at last decided that Burke and King should go up the creek and endeavour to find the main camp of the natives and obtain food from them. Wills, who was now so weak as to be unable to move, was left lying under some boughs, with an eight days' supply of nardoo and water, the others trusting that within that period they would have returned to him.

On the 26th of June the two men started, and poor Wills was left to meet death alone. By the entries in his diary, which he kept written up as long as his strength remained, he evidently retained consciousness almost to the last. So exhausted was he that death must have come to him as a merciful release from the pain of living. His last entries, although giving evidence of fading faculties, are almost cheerful. He jocularly alludes to himself as Micawber, waiting for something to turn up. But it is evident that he had given up hope, and was waiting for death's approach, calm and resigned, without fear, like a good and gallant man.

Burke and King did not advance far. On the second day Burke had to give in from sheer weakness; the next morning when his companion looked at him he saw by the breaking light that his leader was dead.

The last entries in Burke's pocket-book run thus:—

"I hope we shall be done justice to. We have fulfilled our task but have been aban———. We have not been

John King.
From a photo in the possession of E. J. Welch.

followed up as we expected, and the depot party abandoned their post. . . King has behaved nobly. He has stayed with me to the last, and placed the pistol in my hand, leaving me lying on the surface as I wished.''

Left to himself, King wandered about in search of the natives, and, not finding them, the lonely man returned to the spot where they had left Wills, and found that his troubles too were over. He covered up the corpse with a little sand, and then left once more in search of the natives. This time he found them, and, moved by his solitary condition, they helped him to live until rescued by Howitt's party on September 15th.

Meanwhile the absence of any news from Wright, in charge of the main body, was beginning to create a feeling of uneasiness in Melbourne. A light party had already been equipped under A. W. Howitt to follow up Burke's tracks, when suddenly despatches from the Darling arrived from Wright, telling of the non-arrival of the four men. Howitt's party was doubled, and he was immediately sent off to Cooper's Creek to commence a search for the missing men. He had not far to go. On the 13th of September he arrived at the fateful depot camp on Cooper's Creek, with Brahe. He immediately commenced to follow, or try to follow, Burke's outward track, but on Sunday the 15th, while still on Cooper's Creek, King was found by E. J. Welch, the second in command of the relief party. Welch's account of the finding of King is as follows:—

Edwin J. Welch second in command of the Howitt Relief Expedition, and the first man to find King.

"After travelling about three miles, my attention was attracted by a number of niggers on the opposite bank of the creek, who shouted loudly as soon as they saw me, and vigorously waved and pointed down the creek. A feeling of something about to happen excited me somewhat, but I little expected what the sequel was to be. Moving cautiously on through the undergrowth which lined the banks of the creek, the blacks kept pace on the opposite side, their cries increasing in volume and intensity; when suddenly rounding a bend I was startled to see a large body of them gathered on a sandy neck in the bed of the creek, between two large waterholes. Immediately they saw me, they too commenced to howl and wave their weapons in the air. I at once pulled up, and considered the propriety of waiting the arrival of the party, for I felt far from satisfied with regard to their intentions. But here, for the first time, my favourite horse—a black cob known in the camp as Piggy, a Murray Downs bred stock-horse of good repute both for foot and temper—appeared to think that his work was cut out for him, and the time had arrived in which to do it. Pawing and snorting at the noise, he suddenly slewed round and headed down the steep bank, through the undergrowth, straight for the crowd as he had been wont to do after many a mob of weaners on his native plains. The blacks drew hurriedly back to the top of the opposite bank, shouting and gesticulating violently, and leaving one solitary figure apparently covered with some scarecrow rags and part of a hat prominently alone in the sand. Before I could pull up I had passed it, and as I passed it tottered, threw up its hands in the attitude of prayer and fell on the sand. The heavy sand helped me to conquer Piggy on the level, and when I turned back, the figure had partially risen.

"Hastily dismounting, I was soon beside it, excitedly asking:—'Who in the name of wonder are you?' He answered, 'I am King, sir.' For the moment I did not grasp the thought that the object of our search was attained, for King being only one of the undistinguished members of the party, his name was unfamiliar to me.

" 'King,' I repeated. 'Yes,' he said; 'the last man of the exploring expedition.' 'What! Burke's?' 'Yes,' he said. 'Where is he—and Wills?' 'Dead, both dead, long ago,' and again he fell to the ground.

"Then I knew who stood before me. Jumping into the saddle and riding up the bank, I fired two or three revolver shots to attract the attention of the party, and on their coming up, sent the other black boy to cut Howitt's track and bring him back to camp. We then put up a tent to shelter the rescued man, and by degrees we got from him the sad story of the death of his leader. We got it at intervals only, between the long rests which his exhausted condition compelled him to take."

As soon as King had recovered enough strength to accompany the party, they went to the place where Wills had breathed his last; and found his body in the gunyah as King had described it. There it was buried. On the 21st Burke's body was found up the creek; he too was at first buried where he died. Howitt, after rewarding the blacks who had cared for King, started back for Melbourne by easy stages. On his arrival there he was sent back to disinter the remains of the dead; a task which he and Welch safely accomplished, bringing the bodies down by way of Adelaide.

Dr. Beckler, Stone, Purcell, and Patton were the others whose lives were sacrificed on this expedition, so marked with disaster. These victims received no token of public recognition of their fate, although a public funeral was accorded to Burke and Wills, and a statue has been erected to their memory in Melbourne.

The foolish and unaccountable oversight of Burke and his companions in not marking a tree, or otherwise leaving some recognisable sign of their return at the depot, seems to have led Brahe astray completely. He states his side of the case as follows:—

"Mr. Burke's return being so soon after my departure caused the tracks of his camels to correspond in the character of age exactly with our own tracks. The remains of three separate fires led us to suppose that

The Burke and Wills Statue, Melbourne.

blacks had been camped there. . . The ground above the cache was so perfectly restored to the appearance it presented when I left it, that in the absence of any fresh sign or mark of any description to be seen near, it was impossible to suppose that it had been disturbed."

The story of the lost explorers created intense excitement throughout the other colonies. Queensland, as the colony wherein the explorers were supposed to have met with disaster, sent out two search parties. The "Victoria," a steam sloop, was sent up to the mouth of the Albert River in the Gulf of Carpentaria, having on board William Landsborough, with George Bourne as second in command, and a small and efficient party; another Queensland expedition, under Fred Walker, left the furthest station in the Rockhampton district; and from South Australia John McKinlay started to traverse the continent on much the same line of route as that taken by the unhappy men.

Chapter XV.

THE RELIEF EXPEDITIONS AND ATTEMPTS TOWARDS PERTH.

(i.)—John McKinlay.

John McKinlay was born at Sandbank, on the Clyde, in 1819. He first came to the colony of New South Wales in 1836, and joined his uncle, a prosperous grazier, under whose guidance he soon became a good bushman with an ardent love of bush life. He took up several runs near the South Australian border, and thenceforth became associated with that province.

In 1861 he was appointed leader of the South Australian relief party, and started from Adelaide on October 26th. On arriving at Blanche Water, he heard a vague rumour from the blacks that white men and camels had been seen at a distant inland water; but put little faith in the story. He traversed Lake Torrens, and, striking north, crossed the lower end of Cooper's Creek at a point where the main watercourse is lost in a maze of channels. Here he learned definite and particular details respecting the rumoured white men, and thinking there might be some groundwork of truth in the report, he now pressed forward to the locality indicated. Having formed a depot camp, he went ahead with two white men and a native. Passing through a belt of country with numerous small shallow lakelets, they came to a watercourse whereon they found signs of a grave, and they picked up a battered pint-pot. Next morning, feeling sure that the ground had been disturbed with a spade, they opened what proved to be a grave, and in it found the body of a European, the skull marked, so McKinlay states, with two sabre cuts. He noted down the description of the body, the locality, and its surroundings; and in view of these particulars, it has been stated that the body was that of Gray, who died in the neighbourhood.*

See page 192.

Considering the minute and circumstantial accounts that have from time to time been related by the blacks concerning Leichhardt, one is not astonished at the legends told to McKinlay. The native with him told him that the whites had been attacked in their camp, and that the whole of them had been murdered; the blacks having finished by eating the bodies of the other men, and burying the journals, saddles, and similar portions of the equipment beside a lake a short distance away. A further search revealed another grave—empty—and there were other and slighter indications that white men had visited the neighbourhood, so that McKinlay was led to place some credence in this story.

Next morning a tribe of blacks appeared; and although they immediately ran away on perceiving the party, one was captured who corroborated the statement made by the other native. Both of them bore marks on them like bullet and shot wounds. The second native said that there was a pistol concealed near a neighbouring lake. He was sent to fetch it; but returned the next morning at the head of a host of aboriginals, armed, painted, and evidently bent on mischief. The leader was obliged to order his men to fire upon them, and it was only after two or three volleys that they retired.

McKinlay was now satisfied that he had discovered all there was to find of the Victorian expedition, and, after burying a letter for the benefit of any after-comers, he left Lake Massacre, as it was mistakenly named, and returned to the depot camp. His letter was as follows:—

"S.A.B.R. Expedition,
October 23rd, 1861.

To the leader of any expedition seeking tidings of Burke and party:—Sir, I reached this water on the 19th instant, and by means of a native guide discovered a European camp, one mile north on west side of flat. At or near this camp, traces of horses, camels, and whites were found. Hair, apparently belonging to Mr. Wills, Charles Gray, Mr. Burke, or King, was picked up from the surface of a grave dug by a spade, and from the skull of a European buried by the natives. Other less important traces—such as a pannikin, oil-can, saddle-stuffing, &c., have been found. Beware of the natives, on whom we have had to fire. We do not intend to return to Adelaide, but proceed to west of north. From information, all Burke's party were killed and eaten.

JNO. MCKINLAY.

P.S.—All the party in good health.

"If you had any difficulty in reaching this spot, and wish to return to Adelaide by a more practicable route, you may do so for at least three months to come by driving west eighteen miles, then south of west, cutting our dray track within thirty miles. Abundance of water and feed at easy stages."

McKinlay next sent one of his party—Hodgkinson—with men and pack-horses to Blanche Water, to carry down the news of his discovery, and to bring back rations for a prolonged exploration. Meanwhile he remained in camp. From one old native with whom he had a long conversation, he obtained another version of the alleged massacre, in which there was apparently some vestige of truth.

The new version was to the effect that the whites, on their return had been attacked by the natives, but had repulsed them. One white man had been killed, and had been buried after the fight, whilst the other whites went south. The natives had then dug up the body and eaten the flesh. The old fellow also described minutely the different waters passed by Burke, and the way in which the men subsisted on the seeds of the nardoo plant, all of which he must have heard from other natives.

After waiting a month, Hodgkinson returned, bringing the news of the rescue of King and the fate of Burke and Wills. This explained McKinlay's discovery as that of Gray's body, the narrative of the fight and massacre being merely ornamental additions by the natives. After an easterly excursion, in which he visited the two graves on Cooper's Creek, McKinlay started definitely north. It is difficult to follow without a map the Journal containing the record of his travel during the first weeks. Not only does he give the native name of every small lakelet and waterhole in full, but he omits to give the bearing of his daily course.

A northerly course was however, in the main pursued, and McKinlay describes the country crossed as first-class pastoral land. As it was then the dry season of the year, immediately preceding the rains, it proves what an abnormally severe season must have been encountered by Sturt when that explorer was turned back on his last trip in much the same latitude. On the 27th of February, the wet season of the tropics set in; but fortunately the party found a refuge among some stony hills and sand-ridges, in the neighbourhood of which they were camped, though at one time they were completely surrounded by water. On March 10th, the rain had abated sufficiently to allow them to resume their journey; but the main creek which they still continued to follow up north, was so boggy and swollen that they were forced to keep some distance from its banks. This river, which McKinlay called the Mueller, is one of the main rivers of Central Australia, and an important affluent of Lake Eyre, and is now known as the Diamantina. McKinlay left it at the point where it comes from the north-west, and following up a tributary, he crossed the dividing range, there called the McKinlay Range, in about the same locality as Burke's crossing. He had christened many of the inland watercourses on his way across, but most of his names have been replaced by others, it having been difficult subsequently to identify them. In many cases, the watercourses which he thought to be independent creeks, are but ana-branches of the Diamantina.

Passing through good travelling country, and finding ample grass and water, he reached the Leichhardt River flowing into the Gulf of Carpentaria, on the 6th of May.

As his rations were becoming perilously low, McKinlay was anxious to get to the mouth of the Albert, it having been understood that Captain Norman, with the steamship "Victoria" was there to form a depot for the use of the Queensland search parties. His attempts to reach it however, were fruitless, as he was continually turned back by mangrove creeks both broad and deep, and by boggy flats; so that on the 21st of May he started for the nearest settled district in North Queensland, in the direction of Port Denison.

He followed much the same route as that taken by A. C. Gregory on his return from the Victoria River.* Crossing on to the head of the Burdekin, he followed that river down, trusting to come across some of the flocks and herds of the advancing settlers. On reaching Mount McConnell, where the two former explorers had crossed the Burdekin, he continued to follow the river, and descended the coast range where it forces its way through a narrow gorge. Here on the Bowen River, he arrived at a temporary station just formed by Phillip Somer, where he received all the accustomed hospitality. Since leaving the Gulf, the explorers had subsisted on little else but horse and camel flesh, and were necessarily in a weak condition. Had they but camped a day or two when on the upper course of the Burdekin, they would have been relieved much earlier, for the pioneer squatters were already there, and the party would have been spared a rough trip through the Burdekin Gorge. In fact the tracks of the camels were seen by one pioneer at least, a few hours after the caravan had passed. E. Cunningham, who had just then formed Burdekin Downs station, tells with much amusement how McKinlay's tracks puzzled him and his black boy. The Burdekin pioneers did not of course, expect McKinlay's advent amongst them, although they knew that he was then somewhere out west; and such an animal as a camel did not enter into their calculations. Cunningham said that the only solution of the problem of the footprints that he could think of was that the tracks were those of a return party who had been looking for new country, and that their horses, having lost their shoes and becoming footsore, they had wrapped their feet in bandages.

* See page 250.

For his services on this expedition, which were of great value in opening up Central Australia, McKinlay was presented with a gold watch by the Royal Geographical Society, and was voted £1,000 by the South Australian Government.

During the early settlement of the Northern Territory, much dissatisfaction had arisen concerning the site chosen at Escape Cliffs. McKinlay was sent north by the South Australian Government to select a more favourable position, and to report generally on the capabilities of the new territory. He organized an expedition at Escape Cliffs, and left with the intention of making a long excursion to the eastward. But a very wet season set in, and he had reached only the East Alligator River when sudden floods cut him off and hemmed him in. The whole party would have been destroyed but for the resourcefulness displayed by the leader, who made coracles of horse-hides stretched on frames of saplings, by which means they escaped. On his return, McKinlay examined the mouth of the Daly River, and recommended Anson Bay as a more suitable site, but his suggestion was not adopted. McKinlay, whose health suffered from the effect of the hardships incident to his journeys, retired to spend his days in the congenial atmosphere of pastoral pursuits, and died, in 1874, at Gawler, South Australia, where a monument is erected to his memory.

(ii.)—William Landsborough.

William Landsborough, the son of a Scotch physician, was born in Ayrshire and educated at Irvine. When he came to Australia, he settled first in the New England district of New South Wales, and thence removed to Queensland. In 1856, his interest in discovery and a desire to find new country, led him to undertake much private exploration, principally on the coastal parts of Queensland, in the district of Broadsound and the Isaacs River. In 1858 he explored the Comet to its head, and in the following year the head waters of the Thomson.

An old friend and erstwhile comrade, writing of him, says:—"Landsborough's enterprise was entirely founded on self-reliance. He had neither Government aid nor capitalists at his back when he achieved his first success as an explorer. He was the very model of a pioneer—courageous, hardy, good-humoured, and kindly. He was an excellent horseman, a most entertaining and, at times, eccentric companion, and he could starve with greater cheerfulness than any man I ever saw or heard of. But, excellent fellow though he was, his very independence of character and success in exploring provoked much ill-will."

Landsborough was recommended for the position of leader by the veteran A. C. Gregory, and on the 14th of August he left Brisbane in the "Firefly," having on board a party of volunteer assistants who had been stirred by the widespread sympathy with the missing men to take an active part in the relief expedition. Unfortunately, those under Landsborough were, with one exception, unacquainted with bush life. The exception was George Bourne, the second in command, an old squatter who had seen and suffered many a long drought, and whose services proved to be of great value. After some mishap the "Firefly," convoyed by the "Victoria," reached the mouth of the Albert River, where the party was safely landed.

After starting from the Albert, Landsborough came unexpectedly upon a river hitherto unknown. It flowed into the Nicholson, and both Leichhardt and Gregory had crossed below the confluence. It was a running stream with much semi-tropical foliage on its banks, running through well-grassed, level country, and he named it the Gregory. As they neared the higher reaches of the Gregory, they found the country of a more arid nature. They ascended the main range, and on the 21st of December, Landsborough found an inland river flowing south, which he named the Herbert. The Queensland authorities subsequently re-christened the stream with the singularly inappropriate name of Georgina. In this river two fine sheets of water were found, and called Lake Frances and Lake Mary. An ineffectual attempt was then made to go westward, but lack of water compelled them to desist.

Landsborough now returned to the depot by way of the Gregory, and, on arriving there, learnt that Walker had been in and had reported having seen the tracks of Burke and Wills on the Flinders. Landsborough thereupon resolved to return by way of the Flinders, instead of going back by boat. They had very little provisions, but by reducing the number of the party, they managed to subsist on short allowance. On this second trip, he followed the Flinders up, and was rewarded by being the first white man to see the beautiful prairie-like country through which it flows. He named the remarkable isolated hills visible from the river Fort Bowen, Mount Brown and Mount Little. From the upper Flinders he struck south, hoping to come across a newly-formed station, but was disappointed, though he saw numerous horse-tracks showing that settlement was near at hand. At last after enduring a long period of semi-starvation, they reached the Warrego, and at the station of Neilson and Williams, first learnt the fate of those whom they had been seeking.

Landsborough was next appointed Resident at Burketown, and afterwards Inspector of Brands for the district of East Moreton. He died in 1886.

(iii.)—P. E. WARBURTON.

Major Warburton was the fourth son of the Rev. Rowland Warburton of Arley Hall, Cheshire, where he was born on the 15th of August, 1813. He was first educated in France. He entered the Royal Navy in 1826, and in 1829 proceeded to Addiscombe College, preparatory to entering the East India Company's service, in which he served from 1831 to 1853, when he retired with the rank of Major. In 1853 he arrived at Albany. From there he went on to Adelaide, and at the end of the same year was appointed Commissioner of

Police, an office which he held until he was placed in charge of the Imperial Pension Department. On his return from his exploring expedition he was voted £1,000 for himself, and £500 for his party. He was created a C.M.G. in 1875, was awarded the Gold Medal of the Royal Geographical Society of London, and he died in 1889.

In 1873 two prominent South Australian colonists, whose names are intimately connected with the promotion of exploration in that colony, Thomas Elder and Walter Hughes, fitted out an expedition which it was hoped would lead to the rapid advancement of geographical knowledge. Unfortunately the result was not commensurate with the ambitious nature of the undertaking. The command was given to Major Warburton, who was instructed to start from the neighbourhood of Central Mount Stuart, and to steer a course direct to Perth. In spite of being provided with a long string of camels, Warburton incurred so much delay in getting through the sandhills that his camels were knocked up and his provisions nearly all consumed before he had advanced half-way. This compelled him to bear up north to the head waters of the Oakover River. Besides the leader, the party consisted of his son Richard; Lewis, a surveyor; one more white man; two Afghans; and a native. Lewis, the surveyor, showed himself to be a most capable man; in fact, but for his energy and forethought, the expedition would have been swallowed up in the sands of the northwest desert.

Major Warburton.

On the 15th of April, 1873, the explorers left Alice Springs and followed the overland line until they reached a creek called Burt's Creek, whence they struck to the westward. After a vain search for the rivers Hugh and Fincke, which were popularly supposed to rise to the north of the McDonnell Ranges, Warburton altered his course to the north-west, meaning to connect with A. C. Gregory's most southerly point on Sturt's Creek. For some distance his way led him through available pastoral country, and in some of the minor ranges beautiful glens were discovered with deep pools of water in their beds. So frightened were the camels by the rocks that surrounded them, that they would not approach them to drink. On the 22nd of May, after travelling for some days in poor sandy country, they came to a good creek with a full head. The whole flat, on to which the creek emerged from the hills, was one vast spring. This place, the best camp they had yet met with, was named Eva Springs. Leaving the main body resting at these springs, the leader, with two companions, started ahead, and was successful in finding some native wells that enabled him to break up his main camp and advance with all the men and material.

On the 5th of June they crossed the boundary-line between the two colonies, and found themselves on the scrubby, sandy tableland common to the interior. At some native wells, which were called Waterloo Wells, they made an enforced sojourn of about a month; in addition they lost three camels, and one of the Afghans nearly died of scurvy. When they were at last enabled to leave the Waterloo Wells, they found themselves plunged into the salt lake country, where the native inhabitants exist on shallow wells and soakage springs. By their reckoning they were now within ten miles of Gregory's Sturt's Creek; but though Warburton made two separate attempts to find the place, he was unable to recognise any country that at all resembled the description given by Gregory. Rightfully ascribing this disappointment to an error in his longitude, he proceeded on a westerly course once more. The tale of each day's journey now becomes a dreary record of travels across a monotonous barren country, and an incessant search for native wells, their only means of sustaining life.

In addition to other causes for delay, the excessive heat caused by radiation from the surrounding sandhills during the day, compelled the leader to spare his camels as much as possible by travelling at night. This naturally led to a most unsatisfactory inspection of the country traversed, and it was impossible to say what clues to water were passed by unwittingly.

Starvation now commenced to press close upon them; the constant delays had so reduced their store of provisions that they were almost at the end of their resources, whilst still surrounded by the endless desert of sand-ridges and spinifex. Sickness, too, befel them, so that almost the full brunt of the work of the expedition was placed upon the capable shoulders of Lewis and the black boy Charley. The time of these two was taken up in watching the smoke of the fires of the natives, or in looking for their tracks. During the early morning and in the evening they could travel a little, but at night the myriad swarms of ants prevented the tired men from obtaining their natural sleep. If they stopped to rest the camels, they only prolonged their own starvation; yet without rest the camels could not carry them ahead in the search for water. On the 9th of October, the camels strayed away during the night, but luckily came across a small waterhole, and at this welcome spot the party rested for a while; indeed with the exception of Lewis and the native, they were all too weak to do aught else. They slaughtered a camel, and were fortunate to shoot a few pigeons and galah parrots, the fresh meat restoring a little of their strength. They had long since despaired of carrying out the original purpose of the expedition. All that they could hope for was to struggle on with the last remaining flicker of life to the nearest settled country. This was the Oakover River, on the north coast, and to the head of the Oakover, therefore, their worn-out camels were directed. They could entertain no hope of relief before reaching the Oakover, for the discoverer of that river, Frank Gregory, a man always reluctant to acknowledge defeat, had been turned from his southward attempt by this very desert across which they were painfully toiling. On the evening that they started for the station, the whole party were about to ride blindly on into waterless country, where, but for the black boy, they would all have perished. The boy had left the camp early in the morning, and, having come across the fresh tracks of some natives, followed them up to their camp, where he found a well. He hastened back to the party to tell them of his discovery, only to find that they had gone. Fortunately he had sharp ears, and hearing the distant receding tinkle of the camel-bell, by dint of energetically pushing on and cooeeing loudly, he managed to attract their attention, and then led them back to the new source of relief. Lewis and the black boy were now the eyes and ears of the party, and but for them the expedition would never have reached the river.

A fresh start was made after a welcome halt at this well. Warburton and his son could scarcely sit their camels, and followed the weary caravan almost with apathy. On the 14th of November Charley found another native well; but its discovery nearly cost him his life. When close to the native camp, he had gone ahead by himself, as he usually did, so as not to startle the aboriginals. The blacks received him kindly and gave him water, but when he cooeed for his companion, they took sudden alarm and attacked him. They had speared him in the arm and back, and cut his head open with a club when Lewis came up just in time to rescue him. Evidently this attack was not premeditated, but caused by the sudden fear aroused by the sight of the white men and camels. At this well Lewis and one of the Afghans went ahead to strike the head of the Oakover, for they thought they must be drawing near the coast, as the nights were growing cool and dewy, and they had found traces of white iron work in an old camp. In a week Lewis returned, having reached a tributary of the river; and on the 5th of December the whole party arrived at the rocky creek that he had found.

They now proceeded slowly down the Oakover, but came across no sign of occupation. The indefatigable Lewis had therefore again to go ahead for help whilst the others waited for him, living on the flesh of the last camel. He had 170 miles to journey over before he reached the cattle station belonging to Grant, Harper, and Anderson, where he was immediately supplied with horses and provisions to take back to the starving men.

It was on the 29th of December as Warburton was lying in the shade thinking moodily that the station must have been abandoned, and that Lewis had surely been compelled to push on to Roebourne, when the black boy from a tree-top gave a cheerful signal. Starting to their feet, the astonished men found the pack-horses and the relief party almost in their camp.

Of the seventeen camels with which they had started, the two that Lewis had taken on to the station were the only survivors; and all their equipment had been abandoned piecemeal in the desert.

(iv.)—William Christie Gosse.

On the 23rd of April, about a week after the departure of Warburton, William Christie Gosse, Deputy Surveyor-General of South Australia, also left Alice Springs on an exploring expedition, having been appointed by the South Australian Government to take charge of the "Central and Western Exploring Expedition." Like Warburton, he was frustrated by dry country in his endeavour to reach Perth. He had with him both white men and Afghan camel drivers, and a mixed outfit of horses and camels. He left the telegraph line and struck westward, soon finding himself in very dry country, where he lost one horse on a dry stage. He made a depot camp on a creek which he called the Warburton, and while on an excursion from this camp he had the singular experience of riding all day through heavy rain and camping at night without water, the sandy soil having quickly absorbed the downpour. On his return he found that the creek at the camp was running, and though repeated attempts had been made by the Afghans to goad one of the camels over, the animal obstinately refused to cross. Probably the leader thought that it was fortunate for the progress of the expedition that they were not likely to meet with many more running streams. After passing both Warburton's tracks and those of Giles, Gosse reached the extreme western point of the Macdonnell Ranges, where another stationary camp was pitched. The leader made a long excursion to the south-west, and at 84 miles, after passing over sand-ridges and spinifex country, caught sight of a remarkable hill, that on a nearer approach proved to be of singular limestone formation.

"When I got clear of the sandhills, and was only two miles distant, and the hill, for the first time coming fairly in view, what was my astonishment to find it was one immense rock rising abruptly from the plain; the holes I had noticed were caused by the water in some places causing immense caves."

William Christie Gosse, Deputy Surveyor-General of South Australia.

This hill, which Gosse made an ineffectual attempt to ascend, he called Ayer's Rock. He returned to his depot camp, crossing an arm of Lake Amadeus as he did so, and moved the main body on to Ayer's Rock. Rain having set in heavily for some days, he pushed some distance into Western Australia, but soon reached the limit of the rainfall. After many attempts to penetrate the sand-hill region which confronted him, the heat and aridity compelled him to turn back.

His homeward course was by way of the Musgrave Ranges, where he found a greater extent of pastoral country than had been thought to exist there. He discovered and christened the Marryat, and followed down the Alberga to within sixty miles of the Overland Line, when he turned north-eastward to the Charlotte Waters station.

Although Gosse's exploration did not add any important new features, he filled in many details in the central map, and was able correctly to lay down the position of some of the discoveries of Ernest Giles.

William Christie Gosse was the son of Dr. Gosse, and was born in 1842 at Hoddesdon in Hertfordshire. He had come to Australia with his father in 1850, and in 1859 had entered the Government service of South Australia. He held various positions in the survey department, and, after his return from the exploring expedition, he was made Deputy Surveyor-General. He died prematurely on August 12th, 1881.

Chapter XVI.

TRAVERSING THE CENTRE.

(i.)—Ernest Giles.

Ernest Giles was born at Bristol, a famous birthplace of adventurous spirits. He was educated at Christ's Hospital, London, and after leaving school came out to South Australia to join his parents, who had preceded him thither. In 1852 he went to the Victorian goldfields, and subsequently became a clerk, first in the Post Office, Melbourne, and afterwards in the county court.

Having resigned his clerkship, he pursued a bush life, and in 1872 made his first effort in the field of exploration. His party was a small one, the funds being found by contributions from S. Carmichael, one of the party, Baron von Mueller, Giles himself, and one of his relatives. The members of the expedition were Giles, Carmichael, and Robinson; 15 horses, and a little dog were included in the equipment. They started from Chambers Pillar, and it was on this journey that Lake Amadeus and Mount Olga were discovered, the two most enduring physical features whose discovery we owe to Giles. The lake is a long, narrow salt-pan of considerable size, but without any important affluents; Mount Olga is a singular mountain situated about 50 miles from the lake. On this trip Giles went over much untrodden country, but the smallness of the party at last convinced him that it was beyond their frugal means to force their way through the desert country to the settlements of West Australia. Giles was fortunate on this his first trip in having two able and willing bushmen for his companions; otherwise he would not have progressed as far as he did and returned in safety. But most untiring endeavours will not compensate for the lack of numbers, and Giles was forced to return beaten from his first attempt.

His second expedition took place about the same time as that undertaken by Gosse. In consequence of a stirring appeal by Baron von Mueller, he had now the advantage of both substantial private help and a small sum from the South Australian Government. The party numbered four: W. H. Tietkins, who afterwards made an honourable name as an independent explorer; the unfortunate Alfred Gibson; and a lad named Andrews, in addition to the leader.

Giles left the settled districts at the Alberga, and made several determined efforts to push through the sandy spinifex desert that had baffled so many. It was during one of these forlorn hopes that Gibson died.

Baron Sir Ferdinand von Mueller

Anxious to reach a range which he had sighted in the distance, and where he hoped to find a change of country, Giles made up his mind to make a determined effort to reach it, carrying a supply of water with him on packhorses. As usual, Tietkins was to accompany him, but as Gibson complained of having been always previously left in camp, he was allowed to go instead. The two kept doggedly on, the horses, as they gave in, being left to find their way back to the main camp. At last, when several days out, they had but two horses left. Giles sent Gibson back on one, with instructions to push on for the camp, taking what little water he wanted out of a keg they had buried on their outward way, leaving the remainder for his use. He himself intended to make a final effort to reach the range.

Giles's horse soon gave in after they parted, and he had to start to return on foot. On his weary way back he saw that one of the abandoned horses had turned off from the trail, and that Gibson's tracks turned off too, seemingly following it. When he reached the keg, he found that the contents were untouched. Fearing greatly that the unfortunate man's fate was sealed, Giles dragged himself on to the camp. A search was at once instituted, but it was fruitless. Neither man nor horse was ever seen again; and the scene of his fate is known as Gibson's Desert.

During his excursions in various directions, Giles discovered and traversed four different ranges of hills. The party were much worried by the hostility of the blacks, and, what with the uneasiness caused by their attacks, the plague of myriads of ants, the loss of Gibson, and the failure of their own hopes, they were forced to return to Adelaide, baffled for a time, but not beaten.

We thus see how the arid belt of the middle country had defied three different explorers—Warburton, Gosse, and Giles—one equipped with camels only, one with camels and horses, and one who had relied on horses alone.

In 1875 Giles took the field once more. This time, owing to the generosity of Sir Thomas Elder, of South Australia, he was well-prepared. He had a fine caravan of camels, and had his former companion Tietkins with him, besides a completely-equipped party.

The start was made from Beltana, the next halting-place being Youldeh, where a depot was formed. From this place they shifted north to a native well, Oaldabinna. As the water supply here proved but scanty, Giles started off to the westward to search for a better place, sending Tietkins to the north on a similar errand accompanied by Young.

Giles pushed his way for 150 miles through scrub and past shallow lakelets of salt water until he came to a native well or dam, containing a small supply of water. Beyond this he went another 30 miles, but finding himself amongst saline swamps and scrub, he then returned to the depot. Tietkins and his companion were not so successful. At their furthest point they had come across a large number of natives, who, after decamping in a terrified manner, returned fully armed and painted for war. No attempts of the two white men to open friendly communication or to obtain any information from them had succeeded.

A slight shower of rain having replenished the well they were camped at, Giles determined to make a bold push to the west, trusting to the powers of endurance of his camels to carry him on to water.

On reaching the dam that he had formerly visited, he was agreeably surprised to find that it had been nearly filled by the late rains. As it now contained plenty of

A Camel Caravan in an Australian Desert.

water for their wants, and there was good feed all around, they rested by it until the supply of water began to show signs of declining.

On the 16th of September, 1875, he left the Boundary Dam, as he called it, and commenced to try conclusions with the desert to the westward. For the first six days of their march the caravan passed through scrubs of oak, mulga, and sandalwood; next they entered upon vast plains well-grassed, with salt-bush and other edible shrubs growing upon them. Crossing these, the camel train again passed through scrub, but not so dense as before.

When 250 miles had been accomplished, Giles distributed amongst the camels the water he had carried with him. As they kept on, sand-ridges began to make their appearance, native smoke was often seen, and they frequently crossed the tracks of the natives.

On the seventeenth day from the Boundary Dam, Tietkins, who judged by the appearance of the sandhills that there was water in the neighbourhood, sent the black boy Tommy on to a ridge lying south of their course. It was fortunate that he did so, for hidden in a hollow surrounded by sandhills was a tiny lake which they were passing by unheeded until Tommy arrested their progress with frantic shouts. Giles gave this place of succour, which he should have named after his companion, the commonplace name of Victoria Spring; and here the caravan rested for nine days.

Recruited and in good spirits, they soon found themselves amongst the distinctive features of the inner slopes of Western Australia—outcrops of granite mounds and boulders, salt lakes, and bogs. Their next camp of relief was at a native well 200 miles from Victoria Spring.

The quietude of their life at this encampment was however rudely broken by the natives. During their stay they had had friendly intercourse with the blacks, but no suspicions of treachery had been aroused. The explorers were just concluding their evening meal when Young saw a mob of armed and painted natives approaching. He caught sight of them in time to give the alarm to the others, who stood to their arms. Giles says in his journal that they were "a perfectly armed and drilled force," though military discipline was a singular characteristic to find amongst the blacks of this barren region. A discharge of firearms from the whites checked their assailants before any spears had been thrown, and probably prevented the massacre of the whole party.

On leaving this camp the caravan travelled through dense scrub, with occasional hills and patches of open country intervening. They were fortunate to find some wells on the way, and on the 4th of November arrived at an outside sheep-station in the settled districts of

Western Australia, and Giles's long-cherished ambition was at last fulfilled.

The result of this trip was satisfactory to Giles, who thus saw his many fruitless, though gallant efforts, at last crowned with success; but the journey had no substantial geographical or economic results. It resembled Warburton's in having been a hasty flight with camels through an unknown country, marking only a thin line on the map of Australia. An explorer with the means at his command, in the shape of camels, of venturing on long dry stages with impunity, is tempted to sacrifice extended exploration of the country bordering his route and the deeper and more valuable knowledge that it brings to rapidity of onward movement. John Forrest, for example, was able, owing to the many minor excursions he was forced to make because of the nature of his equipment, to gain infinitely more knowledge of the geographical details of the country he passed over than either Warburton or Giles.

Giles now retraced his steps to South Australia, following a line to the northward of Forrest's track. He went by way of the Murchison, and crossed over the Gascoyne to the Ashburton, which he followed up to its head. Then striking to the south of east, he cut his former track of 1873 at the Alfred and Marie Range, the range he had so ardently striven to reach when the unfortunate man Gibson died. How futile was the vain attempt that led to Gibson's death he now realised. He finally arrived at the Peake telegraph station. Few watercourses were crossed; the country was suffering under extreme drought; and no discoveries of importance were made.

Giles published a narrative of his explorations entitled "Australia Twice Traversed." He was a gold medallist of the R.G.S. He entered the West Australian Government service on the Coolgardie goldfields, and, on the 13th of November, 1897, died at Coolgardie, West Australia, where the Western Australian Government erected a monument to his memory.

(ii.)—W. H. TIETKINS AND OTHERS.

W. H. Tietkins was born in London on the 30th of August, 1844, and was educated at Christ's Hospital. He arrived in Adelaide in September, 1859, and took to bush life and subsequently survey-work. On the conclusion of his exploring expeditions with Ernest Giles, he engaged in the survey of Yorke's Peninsula for the S.A. Government, and then paid a visit to England. On his return he went to Sydney, and did some survey work for the New South Wales Government into whose service he permanently entered. He is now a Lands Inspector on the South Coast.

After his experiences as second with Ernest Giles, Tietkins took charge, in 1889, of the Central Australian Exploring Expedition. He left Alice Springs on the overland line on the 14th of March to examine the hitherto unknown country to the north and west of Lake Amadeus. Late in the month of May he discovered and named the Kintore Range, to the north-west of Lake Macdonald, and ascended one of the elevations, Mt. Leisler. During the beginning of the next month he practically completed the circuit of Lake Macdonald and discovered the Bonythorn Ranges to the south-east. On his return journey, Tietkins corrected the somewhat exaggerated notion entertained as to the extent of Lake Amadeus, as he passed through sixty miles of country supposed to be contained in its area without seeing a vestige of this natural feature. In after years he surveyed and correctly fixed its location.

In 1874, surveyor Lewis, the gallant and tireless spirit whose indefatigable efforts had pulled the Warburton

W. H. Tietkins, 1878.

expedition out of the fire took charge of an expedition equipped by Sir Thomas Elder to define the many affluents of Lake Eyre. Starting from the overland line, Lewis skirted Lake Eyre to the north, penetrated to Eyre's Creek, traced that stream and the Diamantina into Lake Eyre, and confirmed the opinion that the waters of Cooper's Creek as well as the more westerly streams found their way into that inland sea. J. W. Lewis, afterwards died in Broome, Western Australia.

In 1875 the Queensland Government decided to send out an expedition to ascertain the amount of pastoral country that existed to the westward of the Diamantina River. It was placed in charge of W. O. Hodgkinson, who had occupied a subordinate position in the Burke and Wills expedition. They started from the upper reaches of the Cloncurry and, crossing the main dividing range on to the Diamantina, followed that river down to the southern boundary of Queensland, where it had been named the Everard by Lewis. This portion was now well-known, and the tracks of the pioneers' stock were everywhere visible. From the lower Diamantina, the party went westwards, and, beyond Eyre's Creek, in good pastoral country, came upon a watercourse which was named the Mulligan. This creek Hodgkinson followed up to the north; and, not knowing that he had crossed its head watershed, went on down the Herbert (Georgina) under the impression that he was still on the Mulligan. He was undeceived when he overtook N. Buchanan with cattle, who was then engaged in re-stocking the stations on the Herbert that had been abandoned in the commercial depression of 1872-3. This was the last exploring expedition sent out by the Queensland authorities, the country within the bounds of that colony being by that time all known.

But across the western border, the vacant and unknown country of South Australia attracted many private expeditions to examine it in search of pastoral holdings. Amongst those from Queensland were two brothers named Prout, who, with one man, went out to look

for new grazing lands, and never returned. Many months
afterwards a search-party, under W. J. H. Carr-Boyd,
found some of the horses, and then the remains of one
of the brothers. It was evident, from the fragments of
a diary recovered, that they had pushed far into the dry
region of South Australia, and had met their deaths from
thirst on the return journey. Probably some of the
waters on which they had relied had unexpectedly failed.

In 1878, Nathaniel Buchanan, a veteran pioneer and
overlander of Queensland, made an excursion from the
Queensland border to Tennant's Creek on the overland
telegraph line. Starting from the Ranken, a tributary of
the Georgina, Buchanan struck a westerly course, and
discovering the head of a well-watered creek running
through fine open downs, he followed it down to the
westward for some days. The creek eventually ran out
into dry flats, so Buchanan struck westward to the
telegraph line, which he reached after some hardship, a
little to the south of Tennant's Creek. The creek which
he discovered, and to which Favenc afterwards gave the
name of Buchanan's Creek, was a most important
discovery, affording a practicable stock route to the great
pastoral district lying between the Queensland border
and the overland line.

Frank Scarr, a Queensland surveyor, was the next to
invade this strip of still unknown land. He attempted
to steer a course south of Buchanan's, but was turned
back by the dry belt of country. On this excursion he
also found two of the horses of the ill-fated Prout
brothers. Scarr then made further north, and, with the
assistance of the creek discovered by Buchanan, was
enabled to reach the line. Owing to the severity of the
drought, however, he was unable to extend his researches
any further, and returned safely to Queensland.

In 1878, a project for a railway line on the land-grant
principle, between Brisbane and Port Darwin was
originated in the former city. The proprietor of the
leading Brisbane newspaper, Gresley Lukin, organized
and equipped a party to explore a suitable line of

country, the object being to ascertain the nature and value of the land in the neighbourhood of the proposed line, and the geographical features of the unexplored portion. The leader was Ernest Favenc, who was accompanied by surveyor Briggs, G. Hedley, and a black boy. They left Cork station on the Diamantina, and kept a north-west course through the untraversed country between that river and the Georgina, or Herbert, as it was then called. They then crossed the border into South Australia, and struck the creek which Buchanan had found, and to which the name of Buchanan's Creek was now given. Leaving this creek at the lowest water, the party struck north, and, after finding two large but shallow lakes came, in the midst of most excellent pastoral country, to a fine lagoon which they named the Corella Lagoon. The trees on the banks of this lagoon, which was about four miles long, were at the time of the visit white with myriads of corella parrots; hence the name. Some three hundred natives were assembled at this lagoon to celebrate their tribal rites; but they showed a friendly disposition.

Ernest Favenc.

From the Corella Lagoon the expedition proceeded north and discovered a large creek running from east to west. It proved to be one of the principal creeks of that region, and was named Cresswell Creek; and a permanent lagoon on it was named the Anthony Lagoon. Cresswell Creek was followed down until, like its fellow creek the Buchanan, it too was absorbed in dry, parched flats. The last permanent water on Cresswell Creek was named the Adder Waterholes, on account of the large number of death-adders that were killed there. A dry stage of ninety miles now intervened between the party and the telegraph line, and the first attempt to cross, on a day of terrible heat, resulted in a return to the Adder Camp, three horses having succumbed to the heat, thirst, and the cracked and fissured arid plains. It being the height of the summer season, and no water within a reasonable distance, it was evidently useless to sacrifice any more horses. There was nothing to do, therefore, but to await at the last camp the fall of a kindly thundershower, by means of which they might bridge the dry gap between them and the line.

The long delay exhausted the supply of rations, but by means of birds—ducks and pigeons—horseflesh, and the usual edible bush plants—blue-bush and pigweed—the party fared sufficiently well.

During their detention at this camp, many short excursions were made, and the country traversed was found to be mostly richly grassed downs. Where flooded country was encroached upon, the dry beds of former lakes were found, encircled in all cases with a ring of dead trees.

In January, 1879, the thunderstorms set in, and the party reached Powell's Creek telegraph station in safety.

This expedition opened up a good deal of fine pastoral country, which is now all stocked and settled.

Western Australia was still busy in the field of exploration. In 1876 Adam Johns and Phillip Saunders started from Roebourne and crossed to the overland line in South Australia. Ostensibly theirs was a prospecting expedition; but as the country to the eastward of the Fitzroy River was then unknown, it was an important exploration event. They were unsuccessful in finding gold, but on their arrival at the line they reported having passed through good pastoral country.

There is no doubt that the east and west tracks of the Queensland explorers, and of Alexander Forrest,* did more to throw open that part of Australia to settlement than did the north and south journey of Stuart, more important as that one was from the purely geographical point of view. Stuart led the way across the centre of the continent, but even after the telegraph line was constructed on his route, very little was known of the country to the east and the west.

The South Australian Government had several times made slight attempts to reach the Queensland border, but in 1878, they sent out H. V. Barclay to make a trigonometrical survey of most of the untraversed country between the line and the Queensland boundary. Barclay left Alice Springs, of which station he first fixed the exact geographical position by a series of telegraphic exchanges with the observatory in Adelaide. Barclay had much dry country to contend against, but managed to reach a north point close to Scarr's furthest south. He did not, however, on that occasion, actually arrive at the Queensland border, but explored the territory on the South Australian side. During the conduct of the survey he discovered and named the Jervois Ranges, the spurs of the eastern Macdonnell, and the following tributaries of Lake Eyre—the Hale, the Plenty, the Marshall, and the Arthur Rivers.

In 1883, Favenc, on a private expedition to report on pastoral country, traced the heads of several of the rivers of the Carpentarian Gulf, and in the following year left the north Newcastle Waters to examine and trace the Macarthur River. The river was followed from its source to the sea, and a large extent of valuable pastoral country and several permanent springs found in its valley; a large tributary, the Kilgour, was also discovered and named. These short excursions, and some exploratory trips made by Macphee, east of Daly Waters, may be said to have concluded exploration between the line and the Queensland border.

In 1883, the South Australian Government despatched an expedition in charge of David Lindsay to complete

*See Chapter XIX.

the survey of Arnhem's Land. Lindsay left the Katherine station, and proceeded to Blue Mud Bay. On the way the party had a narrow escape of massacre at the hands of the blacks, who speared four horses, and made an attempt to surprise the camp of the whites. Lindsay had trouble with his horses in the stony, broken tableland that had nearly baffled Leichhardt; and from one misfortune and another, lost a great number of them. In fact, at one time, so rough was the country that he anticipated having to abandon his horses and make his way into the telegraph station on foot. On the whole, however, the country was favourably reported on, particularly with regard to tropical agriculture.

Another journey was undertaken about this time by O'Donnell and Carr-Boyd, who left the Katherine River and pushed across the border into Western Australia. They succeeded in finding a large amount of pastoral country; but no important geographical discoveries were made.

In 1884 H. Stockdale, who had had considerable experience in the southern colonies, and was an old bushman, made an excursion from Cambridge Gulf to the south, through the Kimberley district. Stockdale found well-grassed country with numerous permanently-watered creeks. When he came to the creek which he named Buchanan Creek, he formed a depot. On his return from an expedition to the south with three men, he found that during his absence the men left in charge of it had been hunting kangaroos with the horses instead of allowing them to rest. There were other irregularities as well, and Stockdale found his resources too much reduced, both in horseflesh and rations, to continue the exploration. They started for the telegraph line, but on the way the two men who had been misbehaving requested to be left behind. As they persisted in their wish, there was nothing left but to accede to it. The two men, with as much rations as could be spared, arms, and powder and shot, were then left at their own request on a permanent creek in a country where game could be obtained. Stockdale himself had to undergo some hardship before reaching the Overland Line. Although search was made for the two men, they were never afterwards found.

One little area of country, of no great importance but still untrodden by man yet remained in Central Australia, as a lure to excite the white man's curiosity. This unvisited spot was situated north of latitude 26, and bounded on the west by the Fincke River, on the north by the Plenty and Marshall Rivers and part of the MacDonnell Ranges, and on the west by the Hay River and the Queensland border. An expedition to exploit it was equipped by Ronald MacPherson, and assisted by the South Australian Government with the loan of camels. The leader was Captain V. Barclay, an old South Australian surveyor, whose name has already been mentioned in these pages.

Barclay had been born in Lancashire, at Bury, on the 6th of January, 1845. He had entered the Royal Navy in 1860, and had been severely wounded on board H.M.S. "Illustrious" by a gun breaking loose when at target practice. He had emigrated to Tasmania in the seventies, and in 1877 had been appointed by the South Australian Government to explore the country lying between the line and the Queensland border, a notice of which occurs in the preceding pages.

The party, lightly-equipped to be more effective, was absent from Oodnadatta from July 24th until December 5th, 1904, and in that time accomplished much useful work in the face of great difficulties. On account of the great heat, the expedition had to resort to travelling by night and resting by day. The country was principally high sandy ridges, some so steep that it was not easy to find crossing-places. They had to sacrifice a lot of valuable stores, personal effects, and a valuable collection of native curios, all chiefly on account of the shortness of water.

By this date the whole of the central portion of Australia was known, and the greater part of it mapped; while all the permanently-watered country had been rapidly utilised by the pastoralists.

Part III.

THE WEST.

John Septimus Roe, First Surveyor-General of West Australia.

Chapter XVII.

ROE, GREY, AND GREGORY.

(i.)—Roe and the Pioneers.

Whilst Sturt and kindred bold spirits had been painfully but surely piecing together the geographical puzzle of the south-east corner of the Australian continent, a similar struggle between man and Nature had commenced in the south-west. Here, Nature kept close her secrets with no less pertinacity than in the east; but, though the struggle was just as arduous, the environment was very different. Instead of rearing an unscalable barrier of gloomy mountains, Nature here showed a level front of sullen hostility. Nor did she lure the first explorers inland with a smiling face of welcome once the outworks had been forced, as she had drawn Evans when he reached the head-waters of the Macquarie and Lachlan. Beyond the sources of the western coastal streams, she fought silently for every eastward mile of vantage ground, spreading before the adventurous intruder the salt lake and the arid desert.

As far back as 1791, George Vancouver, a whilom middy of Cook's, discovered and named King George's Sound, when in command of H.M.S. "Discovery." He formally took possession of the adjacent country, and remained there some days, making a careful survey of both the inner and outer harbours.

On the 9th of December, 1826, Sir Ralph Darling, then Governor of New South Wales, sent Major Lockeyer, of the 57th, with a detachment of the 39th, a regiment intimately associated with the early settlement of Australia, to form a settlement at King George's Sound, where they landed on the 25th of December of the same year. This settlement was established in order to forestall the French, who, according to rumour, intended to occupy the harbour and adjacent lands.

On the 17th of January, 1827, Captain James Stirling, of H.M.S. "Success," left Sydney, intending to survey those portions of the west coast unvisited by Lieutenant King, and also to investigate the nature of the country in the neighbourhood of the Swan River with a view to its suitability for settlement. Stirling was accompanied by Charles Fraser, who had considerable experience as adviser upon Australian sites for settlement. Both Stirling and Fraser reported favourably on the Swan River; and the latter waxing enthusiastic on its eligibility, it was decided to found a new colony there.

In 1829, Captain Fremantle of H.M.S. "Challenger," hoisted the British flag at the mouth of the Swan River, and thenceforth the whole of the Australian continent was under British sway. Captain, now Lieutenant-Governor, Stirling arrived a month later in the transport "Parmelia," and the free colony of Western Australia was launched on its varied career.

The names first mentioned in the annals of land exploration in Western Australia, are those of Alexander Collie and Lieutenant William Preston, who together explored the country on the coast between Cockburn Sound and Geographe Bay. This was in November, 1829, and in the following month Dr. J. B. Wilson, who came to the Sound with Captain Barker on the abandonment of Raffles Bay, made an excursion from the Sound and discovered and named the Denmark River.

In a passage in a letter written by R. M. Davis, of the medical staff, to Charles Fraser, the botanist, there is a detailed reference to this trip:—

"Dr. Wilson, who came here with Captain Barker, started in a direction to Swan Port (Swan River) with a party of men, and in eleven days went over at least two hundred miles of ground. He says, without fear of contradiction in future, that there is far greater proportion of good land in this direction than in any other part of Australia that he had been in, and also wood of large growth, with innumerable rivers. He ascended a very high mountain, which he called Mount Lindsay, in honour of the 39th regiment."

On the 22nd of March, 1830, we first hear of the exploring feats of Lieutenant Roe, R.N., the Surveyor-General of the new colony. Captain John Septimus Roe was born in 1797, and entered the navy. He accompanied Captain P. King to explore the north and north-west coasts of Australia, in 1818, and was a member of King's expedition in 1821. He was the first Surveyor-General of Western Australia, and held that position for forty-two years. He is commonly styled the father of western exploration. He died at Perth on May 28th, 1878. Mrs. Roe, who accompanied her husband to Western Australia in 1829, pre-deceased him in 1870.

On the date mentioned in 1830, Roe was in the field exploring in the vicinity of Cape Naturaliste. Afterwards he was active in the country between the head-waters of the Kalgan and Hay Rivers. In 1836, he first tried serious conclusions with the inland country of Western Australia, when he headed an expedition to explore the tableland that lies to the north and east of Perth. The country was dreary and depressing, and, judging from its configuration and natural properties, he was unable to recommend it as a site for settlement or to depict it as the entrance to more pleasant lands beyond. He reached Lake Brown, near the western boundary of the present Yilgarn goldfield; but the only noteworthy features that he perceived were the salt lakes, that are now so well-known throughout Western Australia. In 1839, Roe distinguished himself by rescuing Grey's dismembered party. On the 14th of September, 1848, he started to make an attempt at further discovery to the eastward. He had with him six men, twelve horses, and three months' provisions. Upon leaving the outer settlements, they encountered the same depressing country as before. Having crossed it, they were turned from their course by scrub of exceeding density, which in turn was succeeded by sandy desert plains. Foiled for the time being they made for the south coast, where they recruited their strength at one of the outlying settlements.

On the 18th they started again, and followed up the course of the Pallinup River. They ascended a branch coming from the north-east, and for a time revelled in the spectacle of well-grassed and promising valleys; but they soon again came amongst the scrub and sand plains of the inland desert. Sighting a granite range to the eastward, they made towards it, but the outlook from its summit brought nothing but exceeding disappointment. Fortunately the weather was showery, and the lack of water did not induce such keen anxiety as the total absence of grass. Still pushing to the eastward, they found their difficulties increase at every step. To the perils of travel through dense thickets and over barren, scorching plains, there was now added the risk of death from thirst. It was not until after days of extreme privation that they reached some elevated peaks, where they obtained a little grass and water.

Their course lay now to the south-east, towards the range sighted by Eyre, and named the Russell Range, and there commenced a desperate struggle with the intervening desert.

So weak were the horses and so compact the belts of scrub, that in three days they had traversed only fifty miles. After being four days and three nights without water for the horses, they reached a rugged hill which they named Mount Riley, where they were relieved by a scant supply. Thence it was but fifty miles to the Russell Range, but the journey involved a repetition of the worst sufferings they had endured. The scrub disputed their passage the whole route, being often so dense as to defy the use of the axe, and many long detours had to be made before they reached their goal.

Every hope they had entertained of a change for the better was shattered by an inspection of the country to which they had so laboriously penetrated. The range, destined to be associated with so many subsequent important explorations, was a mass of naked rocks, and from the summit they could see nothing but the interminable scrub thickets, and in the distance the thin blue line of ocean. Fortunately they found a little grass and water, which saved the lives of their animals. They had discovered a coal seam at the mouth of the Murchison River, and now, on their return journey, they found another at the Fitzgerald River. This was Roe's longest and most important expedition, and it placed him in the front rank of Australian explorers.

Amongst the very early explorers who did as good work as the scanty opportunities permitted, was Ensign R. Dale, of the 63rd Regiment, who pushed east of the Darling Range. Bannister, Moore, and Bunbury, are other noteworthy names amongst those of the early discoverers.

(ii.)—Sir George Grey.

In 1837 an expedition in charge of Captain George Grey and Lieutenant Lushington was sent out from England to the Cape of Good Hope. It was under instructions from Lord Glenelg, and was to procure a small vessel at the Cape to convey the party and their stores to the most convenient point in the vicinity of the Prince Regent's River on the coast. Once landed there, the party was to take such a course as would lead them in the direction of the great opening behind Dampier's Land, where they were to make every endeavour to cross to the Swan River.

Sir George Grey.

The schooner "Lynher" was chartered at the Cape, and on the 3rd of December, 1837, the party was landed at Hanover Bay, with large quantities of live stock, stores, seeds, and plants. Whilst the schooner proceeded to Timor for ponies, Grey employed the time in forming a garden, building sheds for the stores, and in exploring the country in the neighbourhood of Hanover Bay. On the 9th of December, he hoisted the British flag and went through the ceremony of taking possession. On the 17th of January, the "Lynher" returned, and nearly a month later Grey and his party, which now numbered twelve, started from the coast with twenty-six half-broken Timor ponies, as baggage-carriers, and some sheep and goats.

The rainy season had now set in, and many of the stock succumbed almost at the outset, whilst their route proved a veritable tangle of steep spurs and deep ravines. On the 11th of February they came into collision with the natives, and Grey was severely wounded in the hip with a spear. When he had recovered sufficiently to be lifted on to one of the ponies, a fresh start was made, and on the 2nd of March his perseverance was rewarded by the discovery of a river which he named the Glenelg. He followed the course of this river upwards, and reported the country as good, being well-grassed and watered. Sometimes his route lay along the river's bank; at other times by keeping to the foot of a sandstone ridge he was enabled to avoid detours around many wearisome bends.

The party continued along the Glenelg for many days, until indeed they were checked by a large tributary coming from the north. As both the river and the tributary were here much swollen, they had to fall back on the range. It was among the recesses of this range that Grey discovered some curious cave paintings of the blacks, in which the aboriginal figures were represented as clothed.

Unable to find a pass through the mountains, and enfeebled by his wound, Grey determined to retrace his steps. As a last resort he sent Lushington some distance ahead, but there was no noticeable change to report in the aspect of the country. Hanover Bay was reached on the 15th of April. The "Lynher" was waiting there at anchor, and H.M.S. "Beagle" was lying in Port George

* A subsequent photograph of these paintings, by Brockman, is reproduced in Chap. XX.

the Fourth, awaiting the return of Captain Stokes, who was away exploring the coast. The party having embarked, the "Lynher" sailed for the Isle of France, where they safely arrived. Thus ended Captain Grey's first expedition, which is interesting chiefly as a proof of the heroic qualities of its members; for the Glenelg River has never invited settlement, and has yet to prove that it possesses any considerable economic value.

During January, 1839, Grey explored the country between the Williams and the Leschenhault, while searching for a settler who had been lost in the bush.

On the 17th of February in the same year, Grey, who had been back endeavouring to persuade Sir James Stirling to assist him in his explorations, was enabled to start on another exploring enterprise. The object of this, his second important expedition, was to examine the undiscovered parts of Shark's Bay, and to make excursions as far inland as circumstances permitted.

Rock Painting, N.W. Australia.

The party comprised four of the members of his first expedition, five other men, and a Western Australian aboriginal, and they left Fremantle in an American whaler, taking three whale-boats with them. They were duly landed at Bernier Island, where their troubles commenced at once. The whaler sailed away, taking with her by mistake the whole of their supply of tobacco. There was no water on the island, and, in their first attempt to start, one of the boats was smashed and nearly half-a-ton of stores lost. The next day they succeeded in making Dorre Island, but that night both

the remaining boats were driven ashore by a violent
storm. Two or three days were spent in making good the
damage, when they succeeded in making the mainland,
and obtained a supply of fresh water. They had landed
at or near the mouth of a stream which afterwards
proved to be the second longest river in Western
Australia. Grey named it the Gascoyne, and found that
it was then dry beyond the limit of tidal influence. They
then pulled up the coast, but one night, when effecting
a landing, both boats were swamped, and their previously-
damaged provisions suffered another soaking. This
accident kept them prisoners for a week till the wind
and surf had abated. Tired, hungry, and ill, they were
here harassed by frequent threats and one actual attack
by the blacks. A slight break in the weather tempted
them forth once more, and having succeeded in righting
the boats, they made for the mouth of the Gascoyne,
where they re-filled their water-beakers. On March
20th they made a desperate effort in the teeth of foul
weather to fetch their depot on Bernier Island. We may
picture their dismay when they found that during their
absence a hurricane had swept the island, and scattered
their cherished stores to the four winds.

Their position was now as desperate as could be
imagined: the southerly winds had set in, and they had
to coast along a surf-beaten shore against a head wind.
Their food was scanty, and they were weak with the
constant toils they had undergone. There was nothing
for it, however, but to put to sea again, and they
succeeded in reaching Gantheaume Bay on the 31st of
March. Fate had not yet spent all her wrath on them,
and in attempting a landing, Grey's boat was dashed to
destruction upon a rock, and the other received such
a buffeting as to place it beyond repair. The only
hope of safety lay in an overland march to Perth,
three hundred miles away, upon their twenty pounds
of damaged flour and one pound of salt pork per man;
and yet, so wearied were they with the unceasing battle
against wind and sea, that they even welcomed this
hazardous prospect as a change for the better.

They had not proceeded far before differences of opinion arose. Grey naturally wished the men to cover the ground as quickly as possible whilst their strength lasted, whilst they favoured slow marches, relieved by frequent rests. Grey, who recognised that in their weakened condition they could not replenish their scanty food supplies from the native game, held firmly to his opinion, and made strenuous efforts to quicken their progress; but the comparative safety of the shore had lulled his followers into a feeling of false security; and after goading them along for a hundred miles, bearing the chief burden of the march and sharing much of his scanty food with the black boy, Grey left them to push onwards, and if possible send them assistance. He took two or three picked men with him, and, after terrible sufferings and privations, reached Perth, whence a rescue party was immediately despatched. This party found only one man, Charles Wood, who by more closely following Grey's instructions, had made better progress than the others. The remaining five could not be found, and at the end of a fortnight the rescuers were forced to return on account of the lack of provisions. Roe immediately left with another party, and, after experiencing trouble in tracking the erratic wanderings of the unfortunates, came upon most of them hopelessly regarding a face of rock that stopped their march along the beach, unable to muster sufficient strength to climb it. They had then been three days without water, having nothing in their canteens but a loathsome substitute.

One of them, Smith, a lad of eighteen who had accompanied the expedition as a volunteer, had died two days before the rescue; his body was recovered and buried in the wilderness. Walker, the surgeon and second in charge, was still absent; but he had voluntarily left the main body and had pushed on for assistance towards Fremantle, which he safely reached.

During these unfortunate expeditions, Grey had shown a generous spirit of self-sacrifice combined with high courage and a fine enthusiasm for geographical discovery. But his lack of experience and his ignorance of the local seasonal conditions counterbalanced these, and explained his failures. Afterwards he became Acting Government Resident at Albany, on King George's Sound, and he was at a critical period Governor of South Australia. But Australia proper saw little of him in his after prime, and his fame was built up elsewhere, in New Zealand and at the Cape of Good Hope.

Grey's reports left doubt as to the precise value of the country he traversed under such trying circumstances, but he is justly credited with the discovery of many rivers on the west coast—the Grey, the Buller, the Chapman, the Greenough, the Arrowsmith, the Hutt, the Bowyer, and those important streams, the Murchison and the Gascoyne.

(iii.)—Augustus C. Gregory.

In 1846 we come upon a name destined to become linked with the history of exploration in most parts of Australia. There were three notable brothers of the name of Gregory; but as their expeditions, at least those of Augustus and Frank, were conducted independently, with the exception of the first, we shall deal with them separately. H. C. Gregory, it is true, associated his work mostly with that of his brother, A. C. Gregory, generally in a subordinate position, but Frank Gregory won nearly equal fame with his brother Augustus as an independent explorer.

A. C. Gregory was the son of Lieutenant J. Gregory of the 78th Highlanders. He was born at Farnsfield, Nottinghamshire, in 1819, and came to Western Australia with his parents in 1829 in the "Lotus," 500 tons, Captain Summerson, the second passenger ship that sailed for Western Australia. Lieutenant Gregory had five sons in all: William, Augustus, Francis, Henry, and James. The "Lotus" reached Fremantle about the 10th of October, 1829. Captain Gregory had been obliged to retire from active service, being incapacitated by serious wounds received at "El Hamed," in Egypt, and held a large grant of land from the Imperial Government in lieu of pension. On this grant, situated not far from Perth, he established a farm, and on that farm Augustus and his brothers received the balance of their education and underwent their course of bush training. Augustus, after his last expedition, was appointed in 1859 Surveyor-General of Queensland, in which colony he settled down later, after retiring from active official life. He had a seat in the Legislative Council, and was a prominent freemason. He was created C.M.G. in 1874, and K.C.M.G. in 1903, and had several honours conferred upon him by the Royal Geographical Society. He died in Brisbane, in 1905.

If we except a short excursion down the Blackwood and Kojonup Rivers, his expedition of 1846, in which he was accompanied both by F. T. and H. C. Gregory, was the first important enterprise undertaken by him. It was in August that his party left Captain Scully's station at Bolgart's Springs, about seventy miles from Perth.

Photo, Freeman, Sydney.
Augustus C. Gregory, 1880.

On leaving the settled districts they at once found themselves in the barren country that was damming back the eastward flow of settlement. Having traversed it, they reached a range of granite hills, and turning more to the northward, they kept along these for the sake of the rain-water to be found in the rock holes. On striking again to the east, they encountered an extensive salt lake, and in attempting to cross an arm of this marsh, their horses were bogged, and extricated only after great labour. The lake was afterwards proved to be of great size, and to hem them in completely to the eastward, whilst, owing to its crescent-like formation, for five days it baffled all their attempts to proceed northwards.

Finally abandoning the lake, which they called Lake Moore, they turned to the westward to examine some of the streams crossed by Grey during his return from Shark's Bay. On the head of one of these rivers, the Irwin, they found a seam of coal.

"Having pitched our tent and tethered our horses, we commenced to collect specimens of the various strata, and succeeded in cutting out five or six hundredweight of coal with the tomahawk, and in a short time had the satisfaction of seeing the first fire of West Australian coal burning cheerfully in front of the camp, this being the first discovery of coal in Western Australia."

The party then returned by way of the Moore River to Bolgart Springs, which they reached on the 22nd of September.

The discovery of coal deposits and of country available for settlement was seen to be of great importance by the Government, and Lieutenant Helpman, A. C. Gregory, his brother Henry, and Messrs. Irby and Meekleham, in the colonial schooner "Champion," were despatched to procure a quantity of coal for testing. They were also instructed to make a further inspection of the pastoral capabilities of the district, of which there had been so many conflicting opinions. A three days' examination of the country convinced them that it was suitable for settlement.

In 1846 Gregory took charge of an expedition to the north of Perth, organised by the settlers of the colony, and entitled "The Settlers' Expedition"; its object being to proceed to the Gascoyne River, examining the intervening country as to its suitability for pastoral purposes.

Gregory was accompanied by one of his brothers, Messrs. Burges, Walcott, and Bedart, and private King of the 96th Regiment, of whose services he speaks very highly. This expedition excited great hopes amongst the settlers, who found most of the horses and provisions.

The party left Lefroy's station of Wellbeing on the 9th of September, with ten pack, and two riding-horses, but did not succeed in penetrating any distance beyond the Murchison, being turned back at all points, after repeated efforts, by the belt of impervious scrub between the Murchison and Gascoyne. They therefore returned without seeing the latter river, after having attained a distance of 350 miles from Perth; but they succeeded in finding a considerable extent of available country, both pastoral and agricultural, and in discovering a vein of galena on the Murchison. They re-entered Perth on the 17th of November.

The following month, Gregory, Bland, and three soldiers of the 96th, accompanied Governor Fitzgerald by sea to Champion Bay to examine the new mineral discoveries. The galena lode was found to be more important than had been at first supposed. On their return to the schooner, an affray occurred with the natives, in which the Governor was wounded.

"As the country was covered with dense wattle thickets, the natives took advantage of the ground, and having completely surrounded the party, commenced first to threaten to throw their spears, then to throw stones, and finally one man caught hold of Mr. Bland by the arm, threatening to strike him with a dowak; another native threw a spear at myself, though without effect; but before I could fire at him, the Governor, perceiving that unless some severe example was made, the whole party would be cut off, fired at one of the most forward of our assailants and killed him; two other shots were fired by the soldiers, but the thickness of the bushes prevented our seeing with what effect. A shower of spears, stones, kylies and dowaks followed, and although we moved to a more open spot, the natives were only kept off by firing at any that exposed themselves. At this moment a spear struck the Governor in the leg, just above the knee, with such force as to cause it to protrude two feet on the other side, which was so far fortunate as to enable me to break off the barb and withdraw the shaft. The Governor, notwithstanding his wound, continued to direct the party, and although the natives made many attempts to approach close enough to reach us with their spears, we were able by keeping on the most open ground and checking them by an occasional shot, to avoid their attacks when crossing the gullies.''

The natives followed them for seven miles, but finally desisted, and the whites reached the beach and boarded the "Champion" without further mishap.

In 1856 Gregory made his most celebrated journey in the north of central Australia. An account of this journey might have been included in Part II., but as the name of Gregory is so intimately connected with Western Australia, this section is perhaps the most appropriate place in which to recount its incidents. But its lengthy place in which to recount its incidents. But its numerous details demand another chapter.

CHAPTER XVIII.

A. C. AND F. T. GREGORY.

(i.)—A. C. GREGORY ON STURT'S CREEK AND THE BARCOO.

The Imperial Government having long considered the feasibility of further exploration of the interior of Australia, voted £5000 for the purpose, and offered the command of the expedition to A. C. Gregory. As the inexplicable disappearance of Leichhardt was then exciting much interest in Australia, search for the lost expedition was to form one of its chief duties.

On the 12th of August, 1855, Gregory's party left Moreton Bay in the barque "Monarch," attended by the schooner "Tom Tough." There were eighteen men in all. H. C. Gregory was second in command, Ferdinand von Mueller was botanist, J. S. Wilson geologist, J. R. Elsey surgeon and naturalist, and J. Baines artist and storekeeper. They had on board fifty horses, two hundred sheep, and provisions and stores calculated to last them eighteen months on full rations.

They did not reach Point Pearce, at the mouth of the Victoria River, until the 24th of September. There they separated, the schooner taking the stores up the river, and the "Monarch" proceeding on her voyage to Singapore. The horses had been landed at Point Pearce, whence Gregory, his brother, and seven men took them on overland by easy stages. One night the horses were attacked by crocodiles, and three of them were severely wounded. They followed up the course of the Fitzmaurice River and then passed over rough country, not reaching the Victoria until the 17th. On the 20th they rejoined the members who had gone round by the schooner, and learned that she was aground in the river. A large part of their stores was spoiled; and the number of the sheep had also been reduced to forty, in consequence of their being foolishly kept penned up on board. These losses and accidents considerably weakened Gregory's resources, and it was not until the 24th of November that any excursion on horseback was undertaken. An attempt had previously been made to ascend the river in the portable boat with which the expedition had been supplied, but it was not successful, as the boat could not navigate the rocky bars in safety.

Gregory left camp accompanied by his brother, Dr. von Mueller, and Wilson, taking seven horses and twenty days' rations, his object being to examine the country through which the exploring party would have to travel on their route to the interior. On this preliminary trip, he penetrated as far as latitude 16½ south, whence, finding the tributaries flowing from fine open plains and level forest country, all well-grassed, he returned to the main camp.

On the 4th of January, 1856, Gregory started with a much larger party on an energetic dash into the interior. He had with him six men besides his brother, Dr. von Mueller and Baines the artist, and thirty-six horses. He retraced his steps along his preliminary route, and on the 30th of January, thinking it wise judging from the rapid evaporation of the waterholes, to make his means of retreat secure, he formed a temporary camp, leaving there four men and all the horses but eleven to await his return, whilst he, his brother, Dr. Mueller, and a man named Dean, rode ahead to challenge the desert to the south. On the 9th of February, having run the Victoria out, he crossed an almost level watershed, and found himself on the confines of the desert. From a slight rise he looked southwards:—

"The horizon was unbroken; all appeared one slightly undulating plain, with just sufficient triodia and bushes growing on it to hide the red sand when viewed at a distance."

Gregory reviewed the problem from a logical standpoint. He decided to follow the northern limit of the desert to the westward, until he should find a southern-flowing watercourse which would afford him the opportunity to make a dash beyond its confines.

On the 15th of February he came to a small flat which gradually developed into a channel and ultimately became a creek, running first west, and then south-west. This gave him his desired opening, and he pursued the course of the creek through good open country, finding the water plentiful, though shallow. On February 20th, however, the channel of the creek was lost in an immense grassy plain. The country to the south being sandy and unpromising, Gregory kept westwards, and succeeded in again picking up the channel, now finding the water in it to be slightly brackish. That day he crossed the boundary of Western Australia. The creek now gave promise of continuity, the water-holes taking on a more permanent appearance. It was now pursuing a general south-west course, and Gregory, though still rightly anticipating that it would eventually be lost in the dry interior, determined to follow it as far south as should be compatible with safety. He named the creek Sturt's Creek, after the gallant explorer of that name, who was naturally then often in his mind. The creek maintained its southern course, until, on the 8th of March, it ran out into a mud plain and a salt lake.

"Thus, after having followed Sturt's Creek for nearly 300 miles, we have been disappointed in our hope that it would lead to some important outlet to the waters of the Australian interior; it has, however, enabled us to penetrate far into the level tract of country which may be termed the Great Australian Desert."

Gregory, convinced that no useful results could arise from any attempt to penetrate the inhospitable region to the south, determined to return before the rapidly-evaporating water on which they were dependent should vanish and cut off all retreat. He therefore retraced his steps up Sturt's Creek, and on the 28th of March arrived at his temporary depot, where he found the men all well and the horses much improved in condition.

On the 2nd of April, A. C. Gregory, taking his brother Henry, Baines, and one man, started on an excursion to examine the eastern tributaries of the Victoria, and was absent a little over a fortnight. On their return, the whole of the members started for the landing-place on the Victoria, which they reached on the 9th of May. After all arrangements and preparations had been completed, Gregory, with most of the party, started on the return journey overland to Moreton Bay. The "Tom Tough," now caulked and repaired, was to make her way to the Albert River in the Gulf of Carpentaria, where they would again probably meet.

Traversing the tributaries of the Victoria on his homeward way, Gregory met with no remarkable incident until his arrival on the Elsey, a tributary of the Roper River, which he named after the surgeon of the expedition. It was here that he came upon the last authentic trace of Leichhardt. He describes his discovery as follows:—

"There was also the remains of a hut and the ashes of a large fire, indicating that there had been a party camped there for several weeks; several trees from six to eight inches in diameter had been cut down with iron axes in fair condition, and the hut built by cutting notches in standing trees and resting a large pole therein for a ridge; this hut had been burnt apparently by the subsequent bush fires, and only some pieces of the thickest timber remained unconsumed. Search was made for marked trees, but none found, nor were there any fragments of leather, iron, or other equipment of an exploring party, or of any bones of animals other than those common to Australia. Had an exploring party been destroyed here, there would most likely have been some indications, and it may therefore be inferred that the party proceeded on its journey. It could not have been a camp of Leichhardt's in 1845, as it is 100 miles south of his route to Port Essington; and it was only six or seven years old, judging by the growth of the trees; having subsequently seen some of Leichhardt's camps on the Burdekin, Mackenzie and Barcoo Rivers, a great similarity was observed in regard to the manner of building the hut and its relative position with regard to the fire and water supply, and the position in regard to the great features of the country was exactly where a party going westward would first receive a check from the waterless tableland between the Roper and Victoria Rivers, and would probably camp and reconnoitre ahead before attempting to cross to the north-west coast."

From the Roper the party travelled around the shore of the Gulf, keeping rather more inland than Leichhardt had done. On reaching the Albert they found that the "Tom Tough" had not yet arrived at the rendezvous; and Gregory, leaving a marked tree with a message indicating the situation of some instructions he had buried, pushed onwards.

His route from the Albert lay along much the same line of country as that followed by Leichhardt during his journey to Port Essington. He did not, however, make such a wide sweep to the north, up to the Mitchell, but struck away from Carpentaria at the Gilbert River. He corrected the error Leichhardt had fallen into over the situation of the Albert, and re-named the river that he had mistaken the Leichhardt. The exploring party reached the settled districts at Hay's station, Rannes, south of the Fitzroy; and thence reached Brisbane on the 16th of December, 1856.

To advance the search after Leichhardt, the interest in whose fate had been stimulated by the discovery made by Gregory, a public meeting was held in September, 1857, at which resolutions were passed requesting monetary assistance from the Government, and offering the leadership of a new expedition to A. C. Gregory. The appeal was successful, and accordingly in March, 1858, Gregory left Euroomba station on the Dawson with a party of nine in all, one of his brothers going as second. The expedition was equipped for light travelling, taking as means of carriage pack-horses only, of which there were thirty-one, as well as nine saddle-horses.

Gregory crossed the Nive on to the Barcoo, which he proceeded to run down, finding the country in a very different condition from that in which it bloomed when Mitchell rode rejoicingly along what he thought was a Gulf river. A sharp look out was of course kept for any trace of the missing party, and on the 21st of April they came across another marked tree.

"We discovered a Moreton Bay ash (Eucalyptus sp.), about two feet in diameter, marked with the letter L on the east side, cut through the bark about four feet from the ground, and near it the stumps of some small trees that had been cut with a sharp axe, also a deep notch cut in the side of a sloping tree, apparently to support the ridge-pole of a tent, or some similar purpose; all indicating that a camp had been established here by Leichhardt's party. . . No other indications having been found, we continued the search down the river, examining every likely spot for marked trees, but without success."

Approaching the Thomson River, they found the country suffering from drought although the river was running in consequence of some late rains. As winter was now approaching, there was however no spring in the vegetation, and their horses were suffering great hardship. On the 15th of May they found themselves beyond the rainfall, and realised that lack of water was likely to be added to an absence of grass.

"We, however, succeeded in reaching latitude 23 degrees 47 minutes, when the absence of water and grass—the rain not having extended so far north, and the channels of the river separating into small gullies and spreading on to the wide plains—precluded our progressing further to the north or west; and the only chance of saving our horses was to return south as quickly as possible. This was a most severe disappointment, as we had just reached that part of the country through which Leichhardt most probably travelled if the season was sufficiently wet to render it practicable. Thus compelled to abandon the principal object of the expedition, only two courses remained open—either to return to the head of the Victoria (Barcoo) River and attempt a northern course by the valley of the Belyando, or to follow down the river and ascertain whether it flowed into Cooper's Creek or the Darling.''

The latter alternative was chosen, and they proceeded to retrace their steps down the Thomson, and on reaching the junction of the Barcoo they continued south and west. In fact, following Kennedy's route, they soon found themselves involved in the same difficulties that had beset that explorer. The river—now Cooper's Creek—broke up into countless channels running through barren, fissured plains. Toiling on through these, varied by an interlude of sandhills, Gregory at last reached a better-grassed land, where his famished horses regained a little strength. He reached Sturt's furthest point, and continued on to the point where Strzelecki's Creek carried off some of the surplus flood waters, and finally lost the many channels amongst the sandhills and flooded plains. He again struck Strzelecki's Creek and traced it as he then thought, into Lake Torrens, but in reality into Lake Blanche, for the salt lake region had not then been properly delimited. He reached Baker's recently-formed station, eight miles beyond Mount Hopeless, and thence he went on to Adelaide.

(ii.)—FRANK T. GREGORY.

It was in Western Australia, in March, 1857, that Frank T. Gregory commenced his career as an independent explorer by taking advantage of a sudden heavy downpour of rain on the upper reaches of the Murchison River, which flooded the dry course of the lower portion where he was then engaged on survey work. Gregory at once seized the opportunity thus afforded of examining the upper reaches of this river, from which former explorers had been driven back by the aridity of the country. Accompanied by his assistant, S. Trigg, he proceeded up the river finding, thanks to the wet season that had preceded him, luxuriant grass and ample supplies of water. In consequence, he had a more pleasing account of the country to bring back than the report based on the thirsty experiences of Austin. So easy did he find the country, that only scarcity of provisions prevented him from pushing on to the long-sought-for Gascoyne River. As it was, he returned after an absence of thirteen days, having completed what the "Perth Gazette" of that time justly described as "one of the most unassuming expeditions, yet important in its results."

It was so far satisfactory, and roused such fresh hopes in the minds of the settlers, that they once more formed bright hopes of what the River Gascoyne might have in store for the successful explorer. For a long time now they had become resigned to the conclusion that their northern pathway was barred by a dry, scrubby country; but they at once took advantage of the promising practical passage along which Frank Gregory had led the way. Another expedition was organised to penetrate to the Gascoyne, and the leadership being naturally offered to Frank Gregory, was accepted by him.

Frank T. Gregory.

On the 16th of April, 1858, he left the Geraldine mine with a lightly-equipped party of six, including J. B. Roe, son of the Surveyor-General. They had with them six pack and six riding-horses, and rations for 60 days.

They proceeded up the Murchison, and on the 25th of the same month they reached a tributary called the Impey, which had been the highest point reached by Gregory the preceding year. This time, however, the party did not find such ample pasture as he had described. Still following the river up until the 30th April, on that day they struck off on a nor'-north-east course, the course of the Murchison tending too much in an easterly direction to lead them speedily on to the Gascoyne. On the 3rd they reached a gentle stony ascent, which proved to be the watershed between the two rivers. Descending the slope to the northward, they soon came to the head of a watercourse flowing northwards. They followed the new creek, and on the 6th of May came to a river joining it from the eastward, which at last proved to be the Gascoyne.

Gregory kept down the south bank of the Gascoyne, and on the 12th of May passed a large tributary coming from the north, which he named the Lyons. On the 17th they ascended a sandy ridge about sixty feet in height, and had a view of Shark's Bay.

He returned along the north bank of the river, and having reached the Lyons, followed that river up. On the 3rd of June he ascended the highest mountain yet discovered in Western Australia, which he named Mount Augustus, after his brother. Gregory gives the elevation at 3,480 feet, but Mount Bruce in the Hammersley Range, to the north of it, has since been found to be higher.* From the summit, however, he had an extensive view, and was enabled to sketch in the courses of the various rivers for over twenty miles.

As they had now been out 51 days, and their supply of provisions was approaching the end, the party turned back at Mount Augustus, and struck southwards. On the 8th the Gascoyne was re-crossed at a place where its course lay through flats and ana-branches. On the 10th of June they again came to the Murchison, and followed it down to the Geraldine mine, and finally reached Perth on the 10th of July. This expedition, so fruitful in its results to the pastoral welfare of the colony, cost the settlers only their contributions in horses and rations, and a cash expenditure of forty pounds.

The discovery of so much fresh available country on the Gascoyne River, with the prospect of a new base for exploration in the tropical regions beyond, attracted the

* 3,800 feet.

attention of English capitalists. The American civil war had so depressed the cotton trade that those interested in cotton manufacture were seeking for fresh fields in which to establish the growth of the plant. Frank Gregory was then in London, and advantage was taken of his presence to urge upon the Home Government and the Royal Geographical Society the desirability of fitting out an expedition to proceed direct to the north-west coast of Australia, accompanied by a large body of Asiatic labourers, and all the necessary appliances for the establishment of a colony.

Fortunately this rash and ill-considered scheme was greatly modified under wise advice. Roe, the Surveyor-General of Western Australia, and other gentlemen practically acquainted with the subject, suggested that the country should be explored before the idea of any actual settlement should be entertained. Acting on this advice, the Imperial Government gave a grant of £2,000, to be supplemented by an equal subsidy by the Colonial Treasury.

Gregory therefore obtained a suitable outfit in London for the party, and left for Perth to complete the necessary details. The usual official delays occurred, and the expedition did not leave Fremantle, in the barque "Dolphin," until 23rd April, 1861, nearly two months later than had been arranged. As the rainy season in northern Australia terminates in March, this delay was unfortunate.

Nickol Bay on the north-west coast was the destination, and was safely reached. The work of disembarkation being completed, the exploring party started on the 25th of May, 1861.

Gregory first pursued a western course, as he wished to cut any considerable river discharging into the sea, and coming from the interior.

On the 29th of May they struck the river which was subsequently named the Fortescue. As this river seemed likely to answer their expectations of a passage through the broken range that hemmed them in to the south, they followed it up. A narrow precipitous gorge forced them to leave the river, and, after surmounting a tableland, they steered a course due south to a high range, which, however, they found too rough to surmount. Making back on to a north-east course, they again struck the Fortescue, above the narrow glen which had stopped them. They followed it up once more through good country, occasionally hampered by its course lying between rugged hills; but they finally crossed the range, partly by the aid of the river-bed, and partly through a gap. On the 18th June, they succeeded in completely surmounting the range, and found that to the south the decline was more gradual. The range was named the Hammersley Range. Their horses had suffered considerably, and had lost some of their shoes in the rough hills. From here they kept south meaning to strike the Lyons River, discovered by Frank Gregory during his last trip. On coming to a small tributary which he named the Hardey, he formed a depot camp. Leaving some of the party and the most sore-footed of the horses, he pushed on with three men, Brown, Harding, and Brockman, taking three packhorses and provisions for eight days.

On the 23rd of June they came on a large western-flowing river, which he called the Ashburton, and which has since proved to be the longest river in Western Australia. Having crossed this river, and still pursuing a southerly course, he arrived at a sandstone tableland, and on the 23rd had, as Gregory writes, "at last the satisfaction of observing the bold outlines of Mount Augustus."

He returned to the depot camp on the 29th, and though anxious to follow up the Ashburton to the east, the condition of his horses's feet and the lack of shoes prevented him. During the return journey to Nickol Bay, he ascended Mount Samson, and from the summit obtained an extensive view that embraced every prominent peak within seventy miles, including Mount Bruce to the north, and Mount Augustus to the south, the distance between these two elevations being 124 geographical miles. They crossed the Hammersley Range on to the level plains of the Fortescue by means of a far easier pass than that used on the outward journey, and arrived at the Bay on the 19th of July.

On the 31st of July Gregory started on a new expedition to the east. On the 9th of August he came to a river which apparently headed from the direction they desired to explore—namely the south-east. Crossing another river, which they named the Shaw, the explorers, still keeping east and south of east, found, on the 27th of August, a river of some importance running through a large extent of good pastoral and agricultural land. This river was named the De Grey, but as their present object was to push to the south-east, they left its promising banks and proceeded into a hilly country where they soon became involved in deep ravines. After surmounting a rugged tableland, they camped that night at some springs.

The next night, the 29th of August, they came, some time after dark, on to the bank of a wide river lined with the magnificent weeping tea-trees. As three of the horses were tired out, Gregory determind to follow this river up for a day or two, instead of closing with a range of granite hills, capped with horizontal sandstones, which loomed threateningly in their path.

So for two or three days they continued on the Oakover, as he christened the river, and followed its western branch; a tributary of that led them in amongst the ranges, whch were threaded by an easy pass. On the 2nd of September they got through the ranges and emerged upon open sandy plains of great extent, with nothing visible across the vast expanse but low ridges of red drift-sand. Here it was Gregory's lot to experience a test almost equal to one of the grim tramps that had tried Sturt and Eyre.

He camped at a native deserted camp, and the next day failing to find any water ahead, had to return and form a depot. Here he left five of the party with instructions to remain three days and then fall back upon the Oakover. He himself, with Brown and Harding, and six horses, went on to find a passage.

So far he had encountered fewer obstacles, and made more encouraging discoveries than had fallen to the lot of any other Western Australian explorer; but he was now confronted with the stern presence that had daunted the bravest and best in Australia. In front of him lay barren plains, hills of drifted sand, and the ominous red haze of the desert. Let Gregory describe the scene in his own words, as the locality has become historic:—

The three men started on the 6th of September, "steering south-south-east along the ranges, looking for some stream-bed that might lead us through the plains, but I was disappointed to find that they were all lost in the first mile after leaving the hills, and as crossing the numerous ridges of sand proved very fatiguing to the horses, we determined once more to attempt to strike to the eastward between the ridges, which we did for fifteen miles, when our horses again showed signs of failing us, which left us the only alternative of either pushing on at all hazards to a distant range that was just visible to the eastward, where, from the numerous native fires and general depression of the country, there was every reason to think a large river would be found to exist, or to make for some deep rocky gorges in the granite hills ten miles to the south, in which there was every prospect of finding water. In the former case the travelling would be smoothest, but the distance so great that, in the event of our failing to find water, we probably should not succeed in bringing back one of our horses; while in the latter we should have to climb over the sand-ridges which we had already found so fatiguing; this course, however, involved the least amount of risk, and we accordingly struck south four miles and halted for the night.

"7th September. The horses did not look much refreshed by the night's rest; we, however, divided three gallons of water amongst them, and started off early, in the hope of reaching the ranges by noon, but we had not gone three miles when one of the pack-horses that was carrying less than forty pounds weight began to fail, and the load was placed on my saddle-horse; it did not, however, enable him to get on more than a couple of miles further, when we were compelled to abandon him, leaving him under the shade of the only tree we could find, in the hope that we could bring back water to his relief. Finding that it would be many hours before the horses could be got on to the ranges, I started ahead on foot, leaving Brown and Harding to come on gently, while I was to make a signal by fires if successful in finding water. Two hours' heavy toil through the sand, under a broiling sun, brought me to the ranges, where I continued to hunt up one ravine after another until 5 p.m. without success. Twelve hours' almost incessant walking, on a scanty breakfast and without water, with the thermometer over a hundred degrees of Fahrenheit, began to tell upon me severely; so much so that by the time I had tracked up my companions (who had reached the hills by 1 p.m. and were anxiously waiting for me) it was as much as I could do to carry my rifle and accoutrements. The horses were looking truly wretched, and I was convinced that the only chance of saving them, if water was not found, would be by abandoning our pack-saddles, provisions, and everything we could possibly spare, and try and recover them afterwards if practicable. We therefore encamped for the night on the last plot of grass we could find, and proceeded to make arrangements for an early start in the morning. There was still a few pints of water in the kegs, having been very sparing in the use of it; this enabled us to have a little tea and make a small quantity of damper, of which we all stood in much need. Camp 77.

"8th September. At 4 p.m. we were again up, having disposed of our equipments and provisions, except our riding-saddles, instruments, and firearms, by suspending them in the branches of a low tree. We divided a pint of water for our breakfast, and by the first peep of dawn were driving our famished horses at their best speed towards the depot, which was now thirty-two miles distant. For the first eight miles they went on pretty well, but the moment the sun began to have power they flagged greatly, and it was not long before we were obliged to relinquish another horse quite unable to proceed. By 9 a.m. I found that my previous day's march, and the small allowance of food that I had taken, was beginning to have its effects upon me, and that it was probable that I could not reach the depot before next morning, by which time the party left there were to fall back to the Oakover; I therefore directed Brown, who was somewhat fresher than myself, to push on to the camp and bring out fresh horses and water, while Harding and myself would do our best to bring on any straggling horses that could not keep up with him. By dark we succeeded in reaching to within nine miles of the depot, finding unmistakable signs towards evening of the condition to which the horses taken on by Brown were reduced, by the saddles, guns, hobbles, and even bridles, scattered along the line of march, which had been taken off to enable them to get on a few miles further."

Maitland Brown.

Next morning they met Brown within a few miles of the depot coming back to them with water. All the horses but the two which had been left at the remotest point were recovered.

Further on Gregory remarks upon the painful effects produced on the horses by excessive heat and thirst:—

"I cannot omit to remark the singular effects of excessive thirst upon the eyes of the horses; they absolutely sunk into their heads until there was a hollow of sufficient depth to bury the thumb in, and there was an appearance as though the whole of the head had shrunk with them, producing a very unpleasant and ghastly expression."

Gregory was now convinced that the sandy tract before him was not to be crossed with the means at his command, so reluctantly he had to return to the Oakover and follow that river down to its junction with the De Grey. Down the united streams, which now bore the name of the De Grey, the weary explorers travelled through good fertile land, until the coast was reached on the 25th of September. The worn-out state of their horses delayed them greatly in getting across a piece of dry country between the Yule and the Sherlock, where one animal had to be abandoned.

On the 18th of October, they reached Nickol Bay, and were gladly welcomed by the crew of the "Dolphin," who had profitably passed their time in collecting several tons of pearl-shell and a few pearls. On the 23rd the horses and equipment were shipped, and the "Dolphin" sailed for Fremantle.

This journey ended Frank Gregory's active life as an explorer; and it was a noteworthy career which now closed. For the western colony he had thrown open to settlement the vast area of the north-western coastal territory; and after relieving the Murchison from the stigma of barrenness that rested on it, he had discovered and made known all the rivers to the north and east, until the Oakover was reached.

It is singular that Frank Gregory should, like nearly all explorers, have erred greatly in the deductions he drew. When forced to turn back from the country beyond the Oakover, he much laments the fact, because, "not only had we now attained to within a very few miles of the longitude in which, from various geographical data, there are just grounds for believing that a large river may be found to exist draining central Australia; but the character of the country appeared strongly to indicate the vicinity of such a feature."

Of course we now know that no such river drains the centre of Australia. On the contrary, beyond Gregory's eastern limit there occurs a long stretch of coastline unmarked by the mouth of any river. Inland, to the southward, the country even to this day is known as the most hostile and repellant desert in Australia, markedly deficient in continuous watercourses. Providence, then, restrained his footsteps from a land wherein earth and sun seem to unite in hostility against the white intruder. It is a pity that Frank Gregory did not give his undoubted powers of description free scope in his Journal. Now and again he gives them rein; but soon calls a halt, as though alarmed that picturesque language should be found in a scientific, geographical journal. His brother Augustus was unfortunately just as correct and precise.

Frank went to reside in Queensland in 1862, and was nominated to the Legislative Council of that colony in 1874. Before going to Queensland he had acted for some time as Surveyor-General of Western Australia. He was married at Ipswich, Queensland, to the daughter of Alexander Hume. He held office for some time in the McIlwraith Ministry, as Postmaster-General. He was a gold medallist of the Royal Geographical Society, and one of the best of the Australian explorers, as bushman, navigator, surveyor, and scientist. He died at Toowoomba, in 1888, on the 24th of October.

Chapter XIX.

FROM WEST TO EAST.

(i.)—Austin.

By 1854 the gold fever was running high in Australia, and each colony was eager to discover new diggings within its borders. Robert Austin, Assistant Surveyor-General of Western Australia, was instructed to take charge of an inland exploring party to search for pastoral country, and to examine the interior for indications of gold.

He started from the head of the Swan River on a north-easterly course, and on the 16th of July reached a lake, rumours of whose existence had been spread by the blacks, who had called it Cowcowing. The colonists had hoped that it would prove to be a lake of fresh water in the Gascoyne valley, but Cowcowing in reality was a salt marsh, no great distance from the starting-point of Austin's expedition.

The lake was dry and its bed covered with salt incrustations, showing that its waters are undoubtedly saline. Thence Austin made directly north, and passing through repellant country, such as always fell to the lot of the early western explorers in their initial efforts, he directed his course to a distant range of table-topped hills. Here he found both grass and water, and named the highest elevation Mount Kenneth, after Kenneth Brown, a member of his party. Thence he kept a north-east course, traversing stony plains intersected by the dry beds of sandy watercourses. Here the party met with dire misfortune. The horses ate from a patch of poisonous box plant, and nearly all of them were disabled. A few escaped, but the greater number never recovered from the effects of the poison, and fourteen died. Pushing on in the hope of finding a safe place in which to recruit,

Austin found himself so crippled in his means of transit that he had to abandon all but his most necessary stores.

He now made for Shark's Bay, whither a vessel was to be sent to render him assistance or take the party home if required. The course to Shark's Bay led them over country that did not tempt them to linger on the way. On the 21st of September a sad accident occurred. They were then camped at a spring near a cave in the face of a cliff, in which there were some curious native rock-paintings. While resting here, a young man named Charles Farmer accidentally shot himself in the arm, and in spite of the most careful attention the poor fellow died of lockjaw in the most terrible agony. He was buried at the cave-spring camp, and the highest hill in the neighbourhood was christened Mount Farmer. His death and burial reminds one of Sturt's friend Poole, who rests in the east of the continent under the shadow of Mount Poole. Thus two lonely graves in the Australian wilderness are guarded by mountains whose names perpetuate the memory of their occupants. And who could desire a nobler monument than the everlasting hills?

Austin now came to the upper tributaries of the Murchison only to find them waterless. Even the deep cut channel of the Murchison itself was dry. They crossed the river, but beyond it all their efforts to penetrate westward were in vain. They had fought their way to within one hundred miles of Shark's Bay, but they had then been so long without water that further advance meant certain death. Even during the retreat to the Murchison, the lives of the horses were saved only by the accidental discovery of a small native well in a most improbable situation, namely, in the middle of a bare ironstone plain. Their only course now was to fall back on the Murchison, hoping that they would find water at their crossing. Austin pushed on ahead of the main body, and struck the river twenty-five miles below their previous crossing, to make the tantalising discovery that the pools of water on which they had fixed their hopes were hopelessly salt.

A desperate and vain search was made to the southward, during a day of fierce and terrible heat; but on the next day, having made for some small hills they had sighted, they providentially found both water and grass. The whole party rested at this spot, which was gratefully named Mount Welcome.

Nothing daunted by the sufferings he had undergone, Austin now made another attempt to reach Shark's Bay. On the way to the Murchison, they had induced an old native to come with them to point out the watering-places of the blacks. At first he was able to show them one or two that in all probability they would have missed, but after they had crossed the Murchison and proceeded some distance to the westward, the water the native had relied on was found to have disappeared, and it was only after the most acute sufferings from thirst and the loss of some more horses, that they managed to struggle back to Mount Welcome.

Austin's conduct during these terrible marches seems to have bordered on the heroic. Whilst his companions fell away one by one and lay down to die, and the one native of the wilds was cowering weeping under a bush, he toiled on and managed to reach a little well which the blackfellow had formerly shown him. Without resting, he tramped back with water to revive his exhausted companions.

At Mount Welcome they found the water on the point of giving out, and weak and exhausted though they were, an immediate start had to be made to the Geraldine mine, a small settlement having been formed there to work the galena lode discovered by Gregory. That they would ever reach the mine the explorers could not hope; they and their horses were in a state of extreme weakness, the distance to the mine was one hundred and sixty miles, and to the highest point on the Murchison, where Gregory had found water, their first stage was ninety miles. They began their journey at midnight, and by means of forced marches, travelling day and night, they reached Gregory's old camp on the river. Fortunately they had found a small supply of water at one place on the way. From this point the worst of their perils were passed. They followed the river down, obtaining water from springs in the banks, and on the 27th of November arrived at the mine, where they were warmly entertained. Thence they returned to Perth, some by sea and some overland.

Austin's exploration had led to no profitable result. Cowcowing had proved only a saline marsh similar to Lake Moore, the large lake which had haunted Gregory; the upper Murchison was not of a nature to invite further acquaintance or settlement; and the whole of the journey had been a disheartening round of daily struggles with a barren and waterless district, under the fiery sun of the southern summer.

Austin thought that eastward of his limit the country would improve; but subsequent explorations have not substantiated his supposition. He had had singularly hard fortune to contend against. After the serious loss he sustained by the poisoning of his horses, a risk that cannot be effectually warded off by the greatest care, he had been pitted against exceptionally dry country, covered with dense scrub and almost grassless, in which the men and horses must assuredly have lost their lives but for his dauntless and heroic conduct.

Austin afterwards settled in North Queensland, and followed the profession of mining surveyor.

(ii.)—Sir John Forrest.

John Forrest, the explorer who ultimately succeeded in crossing the hitherto impassable desert of the western centre, now made his first essay. An old rumour that the blacks had slain some white men and their horses on a salt lake in the interior was now revived, and gained some credence. A black who stated that he had visited the scene of the incident was interviewed, and Baron von Mueller wrote to the Western Australian Government offering to lead a party thither and ascertain if there was any truth in the report. The Government favourably considered the offer, and made preparations to send out a party. Von Mueller was prevented from taking charge, and the command was given to John Forrest, then a surveyor in the Government service. Forrest was born near Bunbury, Western Australia, on the 22nd of August, 1847, and entered the Survey Department of West Australia in December, 1865.

On the 26th of April, 1869, Forrest left Yarraging, then the furthest station to the eastward. When camped at a native well, visited by Austin thirteen years before, he says that he could still distinctly see the tracks of that explorer's horses. Past this spot he fell in with some natives who told him that a large party of men and horses had died in a locality away to the north, and that a gun belonging to the party was in possession of the natives. On closer examination this story was proved to have its origin in the death of Austin's horses.

Forrest continued his journey to the east, and on the 18th came to a large dry salt lake, which he named Lake Barlee. An attempt to cross this lake resulted in the bogging of the horses, and it was only after strenuous exertions that the horses and packs were once more brought on to hard ground. Lake Barlee was afterwards found to be of considerable size, extending for more than forty miles to the eastward.

John Forrest in 1874.

The native guide Forrest had with him now began to express doubts as to his knowledge of the exact spot at which he saw the remains. After considerable search, Forrest came across a large party of the aborigines of the district. These men, however, proved to be anything but friendly; they threw dowaks at the guide, and advised the whites to go back before they were killed. Next morning they had speech with two of them, who said that the bones were those of horses, some distance to the north; they said they would come to the camp the next day and lead the whites there, but they did not fulfil their promise. No other profitable intercourse with the blacks was possible. One old man howled piteously all the time they were in his company, and another, who had two children with him, gave them to understand most emphatically that he had never heard of any horses having been killed, though some natives had just killed and eaten his own brother.

After vainly searching the district for many days, Forrest determined to utilise the remainder of the time at his disposal by examining the country as far to the eastward as his resources would permit. It was now clear that the story of the white men's remains had originated in the skeletons of the horses that perished during Austin's trip. No matter how circumstantial might be a narration of the blacks, they invariably contradicted themselves the next time they were interrogated, and it was evident that no useful purpose would be served by following them on a foolish errand from place to place. Forrest therefore penetrated some distance east, but was not encouraged by the discovery of any useful country. Nevertheless, he started on a solitary expedition ahead, taking only one black boy and provisions for seven days. He reached a point one hundred miles beyond the camp of the main body, to the eastward of Mount Margaret on the present goldfields. He ascended the highest tree he could find, and found the outlook was dreary and desolate. The country was certainly slightly more open than that hitherto traversed, but it was covered with spinifex, interspersed with an occasional stunted gum-tree. Rough sandstone cliffs were visible about six miles to the north-east, and more to the north appeared a narrow line of samphire flats with gum trees and cypress growing on their edges. Of surface water there was no appearance.

On his homeward route Forrest kept a more northerly and westerly course, and crossed Lake Barlee and examined the northern shore; but he found nothing to induce him to modify the unfavourable opinion pronounced on the country by other explorers. He returned to Perth on the 6th August.

Forrest was next placed at the head of an expedition which was to cross to Adelaide by way of the shores of the Great Australian Bight, along the same ill-omened route followed by Eyre, and never trodden since his remarkable journey. This time the historic cliffs were to be traversed with but slight privation and no bloodshed. Though the information supplied by Eyre was considered to be thoroughly trustworthy, it was recognized that with the scanty means of observation at his command and his famished condition, a few important facts might have escaped his notice, and that if his route were followed by a well-equipped party, the terrors of the region might assume less gigantic proportions.

Forrest's company was to consist of the leader and his brother Alexander, two white men, and two natives, one of whom had accompanied Forrest on his former trip. A coasting schooner, the "Adur," of 30 tons, was to accompany them round the coast, calling at Esperance Bay, Israelite Bay, and Eucla, supplying them with provisions at these depots.

On the 30th of March they left Perth. The first part of the journey to Esperance Bay was through comparatively settled and well-known country, so that no fresh interest attached to it. They arrived at Dempster's station at Esperance a few days before the "Adur" sailed into the Bay, and on the 9th of May, 1870, they started on their next stage to Israelite Bay.

From Esperance Bay to Israelite Bay the journey lacked incident, and it was not until Forrest again parted from his relief boat that he had to encounter the most serious part of his undertaking. He had now to face the line of cliffs which frowned over the Bight, behind which he had, as he knew, little or no chance of finding

SOUTH AUSTRALIA
WESTERN AUSTRALIA
INDIAN OCEAN
Great Australian Bight
Great Victoria Desert
Gibson Desert
Alice Springs Station
Warburton, 1873-4
Overland Telegraph Line
Peake
Telegraph Station
Macumba R.
L. Eyre
Port Augusta
ADELAIDE
Kangaroo I.
St. Vincent Gulf
Spencer Gulf
Gawler Ranges
Fowler B.
Head of the Great Australian Bight
Eucla
Eyre
Forrest 1870
Israelite B.
Esperance B.
Albany
Dec. 1825
Bunbury
C. Leeuwin
PERTH
1829
Swan R.
Geraldton
Arrowsmith R.
Houtman Rocks
Gantheaume B.
Steep Point
Dorre I.
Shark B.
Grey 1839
Gascoyne R.
Murchison R.
Nicholson Ra.
Windich Spr.
Weld Spr.
Mt. Russell
Mt. Margaret
Menzies
Kalgoorlie
L. Barlee
L. Moore
L. Cow Cowing
Forrest 1869
Alexander Spr.
Sutherland Ra.
Fort Mueller
Forrest 1874
Giles 1873
L. Amadeus
Musgrave Ra.
Forrest's Route 1869
Forrest's Route 1870
Forrest's Route 1874
Giles's Route 1873
Grey's Route 1836-7 and 1839
KIMBERLEY
W.A.
King Leopold Range
A. Forrest 1879
Pr. Regent R.
Grey 1836-7
Glenelg R.
Brunswick B.
Hanover B.
Collier B.
Buccaneer Arch.
Swan Point
King Sound
Derby
Fitzroy R.
DAMPIER LAND
Beagle B.
Broome
Roebuck B.
Grey's Route 1836-7
A. Forrest's 1879

water for 150 miles. Having made what arrangements
he could to carry water, he left the last water on the 5th
of April. About a week afterwards he reached the break
in the cliffs, where water could be obtained by digging
in the sandhills. Luckily they had found many small
rock-holes filled with water, which had enabled them to
push steadily on. Forrest says that the cliffs, which
fell perpendicularly to the sea, although grand in the
extreme, were terrible to gaze from:—

"After looking very cautiously over the precipice, we
all ran back, quite terrified by the dreadful view."

While resting and recruiting at the sandhills, he made
an excursion to the north, and after passing through
a fringe of scrub twelve miles deep, he came upon most
beautifully-grassed downs. At fifty miles from the sea
there was nothing visible as far as the eye could reach
but gentle undulating plains of grass and saltbush.
There being no prospects of water, he was forced to turn
back, fortunately finding a few surface pools both on his
outward and homeward way.

On the 24th they started from the sandhills for Eucla,
the last meeting-place appointed with the "Adur."
During this stage he kept to the north of the Hampton
Range, and through a country well-grassed but destitute
of surface water. The party reached Eucla on the 2nd
of July, and found the "Adur" duly awaiting them.
Whilst at Eucla, Forrest, in company with his brother,
made another excursion to the north; he penetrated some
thirty miles inland, and found as before boundless plains,
beautifully-grassed, though destitute of any signs of
water.

After leaving Eucla, the explorers had a distressing
stage to the head of the Great Bight, where they finally
obtained water by digging in the sand. On this stage
the horses suffered more than on any previous one,
having had to travel three days without a drink. From
this point they soon reached the settled districts of South
Australia in safety.

Although this journey of Forrest's cannot strictly be called an exploring expedition, inasmuch as he repeated the journey made under such terrible conditions by Eyre travelling in the opposite direction, yet it is of first-rate importance, inasmuch as, owing to the greater facilities he enjoyed, he was able to pronounce a more final verdict than Eyre was able to give. Forrest found that the gloomy thicket was a fringe confined to the immediate coast-line. On every occasion that he penetrated it, he came on good pastoral land beyond. He writes:—

"The country passed over between longitude 126 degrees 24 minutes and 128 degrees 30 minutes E. as a grazing country far surpasses anything I have ever seen. There is nothing in the settled portion of Western Australia equal to it, either in extent or quality; but the absence of permanent water is a great drawback. . . . The country is very level, with scarcely any undulation, and becomes clearer as you proceed north."

On his arrival in Adelaide he received a hearty welcome, and a similar reception was accorded him on his return to Perth. Unfortunately this expedition destroyed all hope of the existence of any river, the mouth of which might have been crossed unwittingly by Eyre.

We now come to that exploit which gained for Forrest a place in the foremost rank of Australian explorers. The western central desert had long defied the explorers in their attempts to cross its dread confines. But the young West Australian took his men and most of his horses through the very heart of the terrible desert. We have seen how three expeditions had started from the east for the purpose of making this continental traverse, all differently composed—one with the aid of camels only, one with a composite equipment of both horses and camels, and the third with only horses. The successful expedition to be now recorded travelled from west to east, and crossed the desert with horses only.

On the 14th of April, 1874, Forrest left Yuin, then the border of settlement on the Murchison, accompanied by his brother Alexander, two white men, and two natives, to endeavour to cross the unknown stretch of desert country that separated the colonies of eastern Australia from the western settlements. Their route at first lay along the Murchison River, following the upper course, which they found to run through well-grassed country, available for either sheep or cattle. From the crest of the head watershed they had a view of their future travelling-ground to the eastward. It appeared level, with low elevations, but there was a lack of conspicuous hills, which did not promise favourably for water-finding, though good pasture might be obtainable.

For the next few days the party were dependent for water on occasional springs and scanty clay-pans. On the 27th, when following down a creek, they suddenly came upon a fine spring, apparently permanent, which is described by Forrest in his journal as one of the best he had ever seen, both the grass and other herbage around being of fine quality. This place he named Windich Springs, after Tommy Windich, one of the blacks who had now been with Forrest on three expeditions. To the north-west was a fine range of hills, which he named the Carnarvon Range. On leaving this oasis, the explorers found themselves in less attractive country; spinifex and sand became more frequent features of the landscape, and the occasional water-supply became precarious.

On the 2nd of June, Forrest discovered the spring which aided them so greatly in their efforts to cross. This he called Weld Springs, and he describes it as unlimited in supply, clear, fresh, and extending down its gully for over twenty chains. At this relief camp they halted in order to rest the horses.

On the 8th Forrest started on a scouting expedition ahead, taking only a black boy with him. He fully anticipated finding water, for as yet they had not reached a waterless region, and he left instructions for the rest to follow in his tracks in a day's time. He was unfortunate in his selection of a course, for it led them for more than twenty miles over undulating sand-ridges,

Tommy Pierre. Tommy Windich. James Kennedy. James Sweeny.

Alexander Forrest (Second in Command). John Forrest (In Command).

Members of the Exploring Expedition, Geraldton to Adelaide, 1874.

without a sight of any indication of the presence of water. At daybreak, from the top of a low stony rise, he obtained an extensive outlook. Far as he could see to the north and east, nothing was visible but the level unending spinifex; not a watercourse or a hill in sight. Evidently they were trespassing on the edge of the central desert.

Turning back they met the remainder of the party about twenty miles from Weld Springs; and the whole body retreated to their lately deserted camp. After a day's rest, Alexander Forrest and a black boy started to the south-east searching for water. At one o'clock sixty or seventy natives appeared on the brow of the rise overlooking the camp. They were painted and dressed in war costume, and evidently planning an attack. After some consultation they suddenly descended the slope and dashed at the camp. Fortunately the whites were on the alert, and a well-directed volley sent them in headlong retreat to their vantage-point on the brow of the ridge, where they held a fresh council of war. Presently they renewed the assault, but a rifle-shot from Forrest put an end to the skirmish. That evening Alexander and the boy returned, and were much surprised to hear of the adventure with the blacks. They had been over fifty miles from camp and had passed over some well-grassed country but had found no water. As their detention at Weld Springs promised to be indefinite, the party then built a rough shelter of stones in order to ensure themselves some measure of protection against night attacks. When this small defence work was finished, Forrest again reconnoitred ahead for water accompanied by one black boy, and found some clay waterholes, of no great extent, but sufficient for camping purposes. Thither the camp was shifted.

On the 22nd the leader made another search in advance, and in thirty miles came to a fine supply of water, in a gully running through a well-grassed plain whereon there was abundance of good feed for the horses. To the south of this spot there was a small salt lake, which he named Lake Augusta. Another good spring in grassy country was also found. On the 30th of June Forrest made a scouting excursion to the eastward, but experienced ill fortune; for having penetrated as far as possible into the spinifex country, his horses gave out. By the aid of some scanty pools of rain water trapped in some rocks, he succeeded in getting a short distance farther on foot, and in reaching a low range. From its summit he obtained an extensive but depressing view, such as too often greeted the explorer at that time and in that part of Australia. Far away to the north and east, the grey horizon was as level and as uniform as the placid sea; spinifex everywhere, unbroken by ranges or elevations within over thirty miles.

He was now worried and perplexed as to the direction of his future movements. The main party were following up his tracks; but to plunge unthinkingly into such a desert as lay in front of them were sheer madness. Fate relented, however, and after much toilsome search Forrest found a small supply of water, enough for a few days, where he gratefully awaited the approach of his companions.

During the short respite thus accorded them, a diligent search for water was made amongst the low ranges, the only alternative being a retreat of seventy miles. A little more water was found to the south-east, and, as there was coarse rough grass around the well, it helped to prolong their rest and afforded more time for further search. This time Alexander Forrest went ahead, and twenty-five miles further to the eastward found a spring, which was named after him, the Alexander Springs.

Another scouting excursion to the east was likewise fortunate, as far as water was concerned, but the feed for the horses was very poor indeed, and they were suffering greatly. They were now within one hundred miles of Gosse's furthest point west, but that hundred miles was one long line of desert perils. Repeated efforts to traverse it only reduced the little remaining strength in the horses, leading to no discovery of water. But at length a kindly shower filled some rock holes to the northeast of their camp, and after much exertion and hardship they reached the old camp that Giles had named Fort Mueller, and were able to congratulate themselves upon having been the first to bridge the central gap of desert that separated the two colonies.

As the course of Forrest's party from Fort Mueller to the telegraph line was more or less the same as that pursued by Gosse, it is unnecessary to follow the journal to its end. It is enough to state that on Sunday, the 27th of September, the telegraph line was reached at a point some distance to the north of the Peake station. Thus safely concluded an expedition that makes a mark in our geographical history, although it was accompanied by no notable discovery. Central Australia had now been crossed in the same zone that had turned back the explorers from the east, and the fact that Forrest got through, equipped with only the ordinary outfit of horses stamped him as a leader of unusual foresight and judgment.

Forrest's last expedition was rather a survey than a journey of discovery. In 1883, in company with several other surveyors, he landed at Roebuck Bay, and examined a large portion of the Kimberley Division. He proceeded from Roebuck Bay to the Fitzroy River, which his brother had lately explored, and examined the intermediate country as far as St. George's Range, reporting that it consisted mainly of rich elevated grassy plains with abundance of water. He also investigated Cambridge Gulf and the lowest part of the Ord River.

After quitting the field of exploration, John Forrest entered the wider arena of politics, in which his reputation was enhanced. He held the office of Premier of Western Australia continuously for ten years, and he still fills a distinguished position among the public men of federated Australia. He was awarded the Gold Medal of the Royal Geographical Society in 1876, and is now a G.C.M.G. and a Privy Councillor.

(iii.)—ALEXANDER FORREST.

Alexander Forrest was born in 1849, and died in 1901. He accompanied his brother, as we have already noted, in two important expeditions, and in 1871 he took charge of a private expedition to the eastward in search of pastoral country. Owing to a late start, he and his party were compelled to make for the coast when they had reached latitude 31 degrees south, longitude 123 degrees east. This course led them to Mount Ragged, whence, proceeding westerly, they returned to Perth by way of Esperance, having penetrated inland six hundred miles and found a considerable area of good country.

In 1879, Alexander Forrest led an expedition from the De Grey River to the now customary goal, the overland telegraph line of South Australia. He left the De Grey on the 25th of February, and reached Beagle Bay on the 10th of April, the country passed over being like most land in the immediate neighbourhood of the coast, poor and indifferent.

Alexander Forrest.

From Beagle Bay he followed the coast round to the Fitzroy, and proceeded up that river until he encountered a range, which was named the King Leopold Range. Here the party left the Fitzroy, of which river Forrest speaks very highly, and struck north, looking for a pass through the range. It proved to be very rough and precipitous, and when at last they reached the sea, they found themselves in an angle, wedged in between the sea and the range, romantic and picturesque, according to Forrest's description, but quite impassable. Here, too, the natives approached them in threatening numbers, but through the exercise of tact, peace was preserved. On the 22nd of June they attacked one tier of the range, and after a steep climb, which caused the death of one horse, they reached the height of 800 feet and camped. Finding it so hard upon the horses, Forrest left them to rest, and went on foot to discover a road. But he came upon endless rugged zigzags, which so involved and baffled him that he gave it up in despair, and returned. He had now, most reluctantly, to abandon the idea of surmounting the range, and to make for the Fitzroy once more. Following up the Margaret, a tributary of the Fitzroy, he managed to work round the southern end of the range, which still frowned defiance at him, and at last reached the summit, the crest of a tableland, whence he saw before him good grassy hills and plains. Of this country, which he called Nicholson Plains, Forrest speaks most enthusiastically, and doubtless, after the late struggle with the range, it must have appeared a perfect picture of enchantment.

On the 24th they reached a fine river, which was then running strong. They named it the Ord, and followed its course for a time. Thence he continued his way to the line, and on the 18th of August came to the Victoria River. From the Victoria, Forrest had a hard struggle to reach the telegraph line. The rations being nearly exhausted, and one man being very ill, the leader started for Daly Waters station, taking one man with him. After much suffering and privation they at last reached the line, and obtained water at some tanks kept for the use of the line repairers. The absence of a map of the line led Forrest to follow it north, away from Daly Waters, and it was four days before they overtook a repairing party and obtained food.

Alexander Forrest was afterwards for many years a member of the Legislative Council of West Australia, was for six years Mayor of Perth and a C.M.G. He died on the 20th June, 1891. A bronze statue was erected to his memory in Perth, W.A., by his friends.

Chapter XX.

LATER EXPLORATION IN THE WEST.

(i.)—Cambridge Gulf and the Kimberley District.

The futile rush for gold to the Kimberley district had one good result—a better appreciation of its pastoral capabilities, and numerous short expeditions were made in search of grazing country.

Amongst these was one by W. J. O'Donnell and W. Carr-Boyd, who explored an area extending from the overland line in the directon of Roebourne, and were fortunate in finding good country. Later, in 1896, Carr-Boyd, accompanied by a companion named David Breardon, who was afterwards out with David Carnegie, visited the country about the Rawlinson Ranges and penetrated to Forrest's Alexander Spring. His name is also known in connection with exploration in the Northern Territory, and he has made several excursions between the Southern goldfields of West Australia and the South Australian border.

Carr-Boyd and Camel. Photographed at Laverton, W.A., October, 1906.

His experiences were not unlike those of the other explorers; he had to struggle on against heat, thirst, and spinifex, and found occasional tracts of pastoral land destitute of surface water.

In 1884 Harry Stockdale, an experienced bushman, started from Cambridge Gulf in order to investigate the country to the southward, and explore the land in its vicinity.

From the Gulf southward, he traversed well-watered and diversified country till he reached Buchanan's Creek, which must be distinguished from the stream of the same name in the Northern Territory of South Australia.* Having formed a depot there, he hoped to make further explorations, but owing to certain irregularities which had occurred among his followers in his absence on a flying trip, he was compelled to start immediately for his destination on the overland line. A very singular incident happened during this latter part of his journey. Two of the men, named Mulcay and Ashton desired, under the plea of sickness, to be left behind, and resisted every attempt to turn them from their purpose. Stockdale reached the line after suffering great hardship, but the fate of the two abandoned men eluded all subsequent search.

(ii.)—LINDSAY AND THE ELDER EXPLORING EXPEDITION.

In 1891 Sir Thomas Elder of South Australia, who had already done much in the cause of exploration, projected another expedition on a large and most ambitious plan. It was called The Elder Exploring Scientific Expedition, and its main purpose was announced to be the completion of the exploration of Australia. A map was prepared on which a huge extent of the continent was partitioned off into blocks each bearing a distinctive letter, A, B, C, D, etc., quite irrespective of the fact that all these blocks had been partially explored and that some had even been settled.

The leadership of the party was offered to and accepted by David Lindsay, who had already won for himself a name as a capable explorer in South Australia. The second in charge was L. A. Wells. As the expedition was in the main destitute of any striking results, a short synopsis of the journey will satisfy our requirements.

* See page 228.

Shortly after the expedition crossed the border-line between South Australia and West Australia, Mr. Leech one of the responsible officers, was despatched on a fruitless trip northward to search for traces of the ill-fated Gibson, who had perished with Giles some seventeen years previously. The expedition then proceeded *via* Fort Mueller to Mount Squires, where water was obtainable. Thence a south-west course was taken to Queen Victoria's Spring. In latitude 29 degrees, 270 miles south of Mount Squires, the eastern end of a patch of good pastoral country was observed. On reaching the springs they were found to be dry, and all the intended exploration which was to be effected from this base had to be abandoned, the party having to push on to Fraser's Range; and this hasty trip through the desert comprised the only useful work done. Lindsay reported that, when

Photo: Duryea, Adelaide.
Sir Thomas Elder, G.C.M.G.

half-way to the Range, they passed some good country consisting of rich red soil, producing good stock bushes but all exceedingly dry. A belt of country deserving the attention of prospectors was also noted. Having rested some time at the Range, they set out to examine, if possible, the western side of the desert they had just traversed, but lack of water compelled them to take an extreme westerly course to the Murchison by way of Mount Monger, passing through a country covered with miserable thicket on a sandy soil with granite outcrops. On the 1st of January, 1892, they reached their destination, when the majority of the members left the party, and the leader was recalled to Adelaide.

At the termination of the original expedition, or rather before its conclusion was absolutely determined on, L. A. Wells made a flying trip into the district lying between Giles's track of 1876 and Forrest's route of 1874. Starting from his depot at Welbundinum, he completed the examination of what was practically the whole of the still unexplored portion in about six weeks, between the 23rd of February and the 4th of April. During this expedition he travelled 834 miles, discovered some fine ranges and hills, a large extent of pastoral country, some apparently auriferous land, but no water of a permanent kind. The results were indeed very promising, more valuable than those of the original Elder Expedition, and Wells, whose hopes had risen with his success, was intensely disappointed to find on his return that the expedition had been disbanded. Both Lindsay and Wells were natives of South Australia, Lindsay having been born at Goolwa, and Wells at Yallum station in the south-east, which was owned by his father and uncle. Wells joined the Survey Department of South Australia when but eighteen, and at twenty-three was appointed assistant-surveyor to the North Territory Border expedition. On the settlement of the border question he returned to Adelaide, and is now engaged on the Victoria River.

David Lindsay.

(iii.)—WELLS AND CARNEGIE IN THE NORTHERN DESERT.

By this time the gold rush to the southern portion of Western Australia had set in strong, and the country that had so long repelled the pastoral pioneer by its aridity was now overrun with prospectors, their camps supplied with water by condensers at the salt lakes and pools. At first the loss of life was very great; for it was not likely that a district that could be safely traversed only by the hardiest and most experienced bushmen would freely yield its secrets to untried men. Of the many deaths that occurred from thirst, no complete record will ever be available. Some unrecognisable and mummified remains may some day be found amid the untrodden waste; but few have yet been tempted to break in upon the solitude of the dead men of the desert.

As the southern goldfields spread and became thickly-populated, the food supply was an important question, and men's eyes naturally turned to the well-stocked northern stations, from which many cattle were being sent south by steamer. Though the distance overland was not prohibitive, the belt of desert country that intervened, upon which Warburton to his sorrow was the first to venture, forbade the passage of stock. This belt of Sahara extended, roughly speaking, from the eastern border of the colony to the head waters of the western coastal rivers. North and south it lay between the parallels of 19 degrees and 31 degrees south. As yet no daring attempt had been made to traverse its barren confines from south to north. But, to the born explorer, difficulty and danger give an added zest to geographical research; and in the year 1896 two separate expeditions sought to cross this dreadful zone. Both left civilization within a few days of each other. The first to start was known as the Calvert Expedition, from its originator. It was under L. A. Wells, the South Australian surveyor who had been the energetic second of the former Elder Expedition. The other was equipped and led by the Hon. David Carnegie.

Wells formed a depot at a spot well provided with camel feed and water, at some distance to the south-west of Forrest's Lake Augusta, which he found, at that time, dry. Here he left the main part of his caravan to await his return whilst he made a flying trip to the north. He was away from the 10th of August to the 8th of September, during which he found at his furthest point, a distance of two hundred miles, a good native well, which he named Midway Well. On the 14th of September the whole party made a start, and reached Midway Well on the 29th, all well. At Separation Well, another good well a little farther to the north, the party separated, C. F. Wells, a cousin of the leader, and G. L. Jones, intending to travel for about eighty miles in a north-west direction to examine the country, and then to return on a north-east course and rejoin the caravan at Joanna Springs, which had relieved Warburton in his extremity. About thirty miles south of Joanna Springs, where the leader expected the two men to cut his tracks, Wells found his camels suffering terribly from the extreme heat and their labours among the constantly-recurring sand-ridges, whilst the scanty native wells they found were insufficient to give their camels water. When at last they reached the latitude of Joanna Springs they had been obliged to abandon three camels and all their equipment except the actual necessaries.

It was also evident that the longitude of the springs given by Warburton was wrong, for all the country

Photo: Duryea, Adelaide.
L. A. Wells.

around was a sandy desert without the slightest indication of well or spring. To linger in such a spot was to court destruction, and they had to push on to the Fitzroy as fast as their worn-out camels could take them. The reader will remember that Warburton had failed to find A. C. Gregory's most southerly point on Sturt's Creek when looking for it, and it was afterwards proved that Joanna Springs had been charted by him about ten miles to the westward of its true position. On the 7th of November, in the darkness of morning they at last reached the Fitzroy, with the camels just at their last gasp.

On the 16th of December, Wells, accompanied by that veteran pioneer N. Buchanan, formerly of Queensland, started back with an Afghan, a native boy, and eight camels, to look for the two men, who he hoped had succeeded in finding Joanna Springs. He was absent until the 10th of January, 1897, when he was forced to return unsuccessful. At the beginning of April, taking with him his former companions of the expedition, Wells renewed the search, and on the 9th at last succeeded in identifying the Joanna Springs of Warburton. On the 13th some articles belonging to the lost men were found amongst the natives, but he did not at that time find the bodies. He started again with two members of the West Australia police force, Sub-Inspector Ord and Trooper Nicholson, and native trackers. This time they were successful in inducing some natives to guide them to the exact spot where the remains lay amongst the spinifex and sand. The bodies were within six miles of the place where, on the last search expedition, Wells had found articles of equipment with the natives.

G. L. Jones had kept a journal which supplied the clue to the cause of their death.

"He stated in his journal," says Wells, "that they had gone west-north-west for five days after separating from the main party, then travelling a short distance north-east, and that both he and Charles felt the heat terribly and were both unwell. They then returned to the well (Separation Well) after an absence of nine days, rested at the water five days, and then started to follow our tracks northward. Afterwards one of their camels died, which obliged them to walk a great deal, and they became very weak and exhausted by the intense heat. When writing he says that two days previously he attempted to follow their camels, which had strayed, but after walking half-a-mile he felt too weak to proceed and returned with difficulty. There was at that time about two quarts of water remaining to them, and he did not think they could last long after that was finished."

From the above extract from Wells's Journal, it is evident that the unfortunate men lost their lives through a mistake in judgment in returning to Separation Well, the straying away of their camels, and the merciless rays of the desert sun.

The account of this, the first expedition to cross the great sandy desert from south to north, confirms in every particular Warburton's experiences of the difficulties of exploration in that region. The intense heat of the sun, and its radiation from the red sand-ridges, the heat from both sky and earth, render it nearly impossible to travel during day, the only time when a man can perceive those slight indications which may eventually lead him to water. The traveller is therefore compelled to make night-stages, and frequently passes unheeding the very pool or well that would have saved his life. During the night not only are the natural physical features difficult to discern, but the birds, those water-guides of the desert, are sleeping.

As soon as the news that Jones and Wells were missing was wired to Perth, the West Australian Government promptly despatched W. P. Rudall in charge of a search-party, from Braeside station on the Oakover River.

Crossing into the desert country, Rudall, guided by blacks, came upon a camp in which footsteps, supposed to be those of the missing men, were traceable. His camels failing him, the tracks were lost, and he was obliged to return. A second search was likewise fruitless, but rumours brought in by the natives of straying camels, caused a third party to be organised. Rudall this time went south of the head of the Oakover, and penetrated the dry spinifex country below the Tropic. Here the bodies of two men, supposed to have been murdered by the natives, were found, but on further investigation it was decided that the remains were not those of the men they were searching for. On his return Rudall started out on a final trip, and penetrated to a point sixty miles south of Joanna Spring before returning. Though these journeys were not successful in attaining the initial object of their search, they were of great service in gaining much information concerning the hitherto unknown desert. Running easterly into this dry belt, Rudall found a creek, which is now known as the Rudall River.

Four days after Wells had started, the Hon. David Carnegie, fourth son of the ninth Earl of Southesk, born March 23rd, 1871, left an outpost of civilization called Doyle's Well, some fifty miles south of Lake Darlot, intending to cross Warburton's Desert on a north-easterly course, about two hundred miles to the east of the route pursued by surveyor Wells. The objects of this purely private expedition were (1) extension of geographical knowledge; (2) the desire to ascertain if any practicable stock-route existed between Kimberley and Coolgardie; (3) the discovery of patches of auriferous country within the confines of the desert. In the two last objects Carnegie was doomed to disappointment, but as a geographical contribution to our scanty knowledge of north-west Australia, the outcome of his repeated journey was distinctly valuable.

David Wynford Carnegie.

Carnegie started with three white men and a native boy, and for many days passed through country that afforded no water for the camels; of which they had nine. A native was induced to lead them to a singular spring situated in a cavern twenty-five feet underground. Though the water was not easy of access, having to be hauled up by a bucket to the surface, there was an ample supply for the camels, and, as Carnegie considered the well to be permanent, he named it the Empress Spring.

The discovery of this subterranean spring was indeed a godsend, as when they eventually reached Forrest's Alexander Spring they found it dry. A similar experience had befallen W. W. Mills who, after Forrest's exploration, had attempted to take over a mob of camels in Forrest's tracks.

Strangely enough a lagoon of fresh water was found at the foot of the creek in which the spring was situated, and this satisfied their wants. From this sheet, which was named Woodhouse Lagoon, the party kept a nearly northerly course across what Carnegie calls in his book "the great undulating desert of gravel." Over this terrible region of drought and desolation the party made their painful way by the aid of miserable native wells, found with the greatest difficulty, and a few chance patches of parakeelia,* until they were relieved by finding, through the good offices of an aboriginal guide, a beautiful spring, which was named Helena Spring. They were then seven days out from Woodhouse Lagoon, and during the last days of the stage they had been travelling across most distressing parallel sand-ridges.

From Helena Spring Carnegie struggled on, intending to strike the northern settlements at Hall's Creek where there is a small mining township. On the way there, while still in unexplored country, they discovered one more oasis, in a rock hole, which was called Godfrey's Tank, after Godfrey Massie, one of the party. On November 25th, 1896, they congratulated themselves that they were at last clear of the desert and its desolation, having come out on to a well-watered shady river, running

* A ground plant which camels eat, and which assuages their thirst.

towards the northern coast. But a sad accident turned their rejoicing into mourning. Charles Stansmore accidentally slipped on a rock, when out shooting, and his gun going off, he was shot through the heart and died instantly. His friend Carnegie speaks most highly of him, and his sudden death on the threshold of success was a sad blow to the company. Stansmore was the third explorer to lose his life from a gun accident.

At Hall's Creek Carnegie heard of the misfortune that had befallen Wells, in the loss of two of his party, and he at once volunteered his assistance; but as search-parties had already started out, his aid was not required. He therefore rested for a short time before again trying conclusions with the desert on the return journey. Sturt's Creek was by this time occupied and stocked, and the party followed it down until they arrived at its termination in Gregory's Salt Sea. From this point Carnegie kept a southerly course to Lake Macdonald near the South Australian border, passing on his way a striking range which he named the Stansmore Range, after his unfortunate companion. Lake Macdonald was long thought to be a continuation of Lake Amadeus, until the exploration of Tietkins in 1889 proved its isolation. From Lake Macdonald, Carnegie, who had now three horses in his equipment, kept a more south-westerly course towards the Rawlinson Range, the endless sand-dunes still crossing his track in dreary succession. So persistently did they rise across his path that, on one day, eighty-six of them were crossed by the caravan during a progress of eight hours. From the Rawlinson Range they kept on the same south-west course until they struck their outward track at Alexander Spring. A fall of rain fortunately replenished the spring shortly after the arrival of the party. They reached Lake Darlot on the 15th of July, and their desert pilgrimage was ended.

Not only did Carnegie get safely across the dreaded desert, but he returned overland to his starting-point by a different route. He wrote a book, "Spinifex and

Sand," which contains a most interesting account of this journey, as well as a graphic and picturesque description of the physical features of the Great Sandy Desert.

Carnegie died before he had made more than this one contribution to Australian geography. Like the ill-fated Horrocks, he had the explorer's ardent spirit. His restless and adventurous soul ever leading him onward to the frontiers of settlement and the outskirts of civilised life, he fell beneath a shower of poisoned arrows at Lokojo in Nigeria, on the west coast of Africa, on the 27th of November, 1900.

(iv.)—HANN AND BROCKMAN IN THE NORTH-WEST.

The isolation of that remote corner of the continent in which Grey had made his maiden effort at exploration, added to the discouraging and forbidding report brought back by Alexander Forrest of his repulse by the King Leopold Range, had deterred further exploration there. Frank H. Hann, who had been a Queensland pioneer, came over to Derby, and, after one or two tentative excursions into the desert country to the south, had his attention drawn to the unknown country to the north of the King Leopold Range. Hann crossed the range with difficulty; but after examining the country to the north and east on the coast side of the range, he was so well satisfied with its pastoral capabilities that he returned to Derby and applied for a pastoral lease.

Photo: Mathewson, Brisbane.
Frank Hann. Explorer of the N.W., and discoverer of a stock route between S.A. and W.A.

Wishing to make a closer examination of the locality, he returned accompanied by Sub-Inspector Ord. Some of the tributaries of the Fitzroy were traced and named, and an extensive river, which Hann called the Phillips, was afterwards re-named the Hann by the Surveyor-General of Western Australia. One very rugged range could not be surmounted, and had to be skirted to the east, as the only apparent gap was an impassable gorge with precipitous sides, through which the Fitzroy River forced a passage. It was named the Sir John Range. After more good pastoral country was found, the party returned to Derby. Hann afterwards, in 1903, made the first of several trips from Laverton, Western Australia, to Oodnadatta in South Australia. He reported having found a practicable stock-route, of which he was chiefly in search, as far as the Warburton Ranges, and some pastoral land north and west of Elder Creek. Since then he made another journey with the same object in view, but encountered extremely dry weather and underwent many hardships. Hann was born in Wiltshire, in 1846, and came to Victoria with his parents at a very early age. He spent most of his life squatting in North Queensland, where he held several station properties.

In the first year of the present century the Western Australian Government followed up Hann's explorations north of the King Leopold Range, by a larger and better-equipped party instructed to make a thorough examination of the region. It was placed in charge of F. S. Brockman, a Government surveyor, who had with him C. Crossland as second, F. House as naturalist, and Gibbs Maitland as geologist.

Brockman was born in Western Australia in 1857, was educated at Bishop's College, and after a spell in the bush on his father's properties, he joined a Government Survey camp, as cadet. In 1879 he started as surveyor on his own account. From 1882 to 1897 he was employed by the Lands and Survey Department in many parts of Western Australia from Cambridge Gulf in the north to the Great Bight in the south. At the time when he was selected to lead the Kimberley expedition, he was Controller of the Field Survey Staff.

Brockman was most successful in securing full information of this long-secluded region; of its geographical, geological, and botanical details. Many interesting photographs were obtained of the different physical features and of the aborigines and their modes of life; amongst them being views of rock paintings similar to the mysterious scenes noticed by Grey during his first expedition to the Glenelg River.

From a photograph by F. S. Brockman.
Aboriginal Rock Painting on the Glenelg River.

The party left Wyndham on Cambridge Gulf and proceeded first southwards and then to the westward to the Charnley River, which had been discovered by Frank Hann. The tributary waters of the Glenelg and Prince Regent Rivers, and the tidal rivers that flow into Collier and Doubtful Bays were also visited, and Brockman traced the Roe River from its source to its outflow in Prince Frederick Harbour. The Moran River was discovered, and its whole course traced to the mouth in the same inlet. The head waters of the King Edward River were discovered at the watershed; and this river was again met lower down and its course traced to its exit. Portions of the shores of Admiralty Gulf, Vanstittart, and Napier Broome Bay were closely examined with a view to selecting a suitable port for the district. The most important practical result of the expedition was the discovery of an area of six million acres of basaltic pastoral country covered with blue grass, Mitchell and kangaroo grasses, and many varieties of

Typical Australian Explorers of the early Twentieth Century.

what is known as top feed. No auriferous country was found, but some fine specimens of the baobab tree were seen, some of them averaging fifty feet in diameter.

We have now turned the last page of the story of those bold spirits who played no mean part in the making of Australasia by exploring the continent. For nearly a century and a quarter the white man had been restlessly searching out and traversing every square mile of the

land, and now, at the beginning of the twentieth century,
his work is finished. And throughout the long struggle
it had ever been a stubborn conflict between the explorer
and the inert forces of Nature. Through the weary toilsome years of arduous discovery, Man and Nature had
seldom marched side by side as friends and allies. When
Nature posed as the explorer's friend and guide, it was
often only to lure him on with a smiling face to his doom.
From the days when the soldier of King George the Third
went forth with his firelock on his shoulder, computing
the distance he covered by
wearily counting the number
of paces he trudged, to the
day when the modern adven-
turer aloft on his camel
eagerly scans the horizon of
the red desert in search of
the distant smoke of a native
fire, and then patiently
tracks the naked denizen of
the wilderness to his hoarded
rock-hole or scanty spring,
the explorer has ever had to
fight the battle of discovery
unaided by Nature. The
aborigines generally either
feigned ignorance of the
nature of the country, or gave only false clues and
misguiding directions. Even the birds and animals of
the untrodden regions seemed to resent the advance of
civilization, and to delight in leading the footsteps of the
white intruder astray. Hence it was by slow degrees, by
careful study of the work of his predecessors in the
field, and often by heeding the warning conveyed in their
unhappy fate, that the Australian explorer added to the
sum of knowledge of his country, and step by step
unveiled the hidden mysteries of the continent.

Ernest Giles.

INDEX OF NAMES OF PERSONS.

ANDREWS, 217
Ashton, 282
Austin, 254, 264-7, 268

BABBAGE, 169-175
Bagot, Walter, 89
Baines, 247-8
Baker, 253
Bannister, 237
Barallier, 4
Barclay, H. V., 227, 229
Barker, Capt., 234
Barrett, 121
Bass, 5, 13
Baxter, 144-6
Beckler, Dr. H., 189
Beckler, Dr. L., 189
Bedart, 244
Berry, Alex., 44
Binney, 125
Black, Wm., 38
Bladen, F. M., 6
Bland, 245
Blaxland, 7-15, 17
Bonney, 135-6
Boyd, Thos., 49
Bourne, 200, 209
Bowen, Governor, 124
Breardon, 281
Brahe, 192 *et seq.*
Briggs, 225
Brisbane, Governor, 44
Brockman, 238, 257, 292-6
Brown, Kenneth, 264
Brown, Maitland, 257 *et seq.*
Browne, Dr., 156, 161, 164, 166, 170
Buchanan, N., 223, 224, 287
Bunbury, 237
Burgess, 244
Burke, 122, 179, 184, 185, 186-200, 202

CALVERT (Leichhardt), 102
Calvert, 285
Cameron, 93
Campbell (South Australia), 170, 172
Campbell, 152
Carmichael, S., 216
Carnegie, D. W., 281, 285-292
Carpenter, 120
Carr-Boyd, 224, 228, 281
Carron, 114, 119
Cayley, 5, 13
Clarke, A. W. B., 105
Clarke ("The Barber"), 78
Classen, 106
Clayton, 69, 74
Collie, Alex., 234
Collins, Capt., 2
Cowderoy, 125
Cox, 19
Crossland, 293
Cunningham, 123, 205
Cunningham, Allan, 26, 29, 50-58, 77, 79
Cunningham, Richard, 80-2
Currie, Capt., 43

DALE, Ensign, 237
Dalrymple, 123
Darling, Governor, 60, 75, 233
Davis, R. N., 234
Dawes, Lieut., 3, 4
Delisser, 172
Dempster, 270
Dixon, 80
Dobson, Capt., 121
Douglas, 119
Dunn, 115
Dutton, 172

EBDEN, 135
Elder, Sir Thos., 209, 218, 223, 282
Elsey, J. R., 247
Eulah, 129
Evans, G. W., 14, 16-22, 26, 36, 37, 233
Eyre, 88, 89, 135-150, 154, 156, 169, 236, 258, 270

FARMER, Charles, 265
Favenc, Ernest, 224-6, 227
Finch, 79
Finnegan, 40
Fitzgerald, Governor, 245
Flinders, 39, 146
Flood, 163
Forrest, Alex., 227, 270, 274, 276, 279-280, 292
Forrest, Sir John, 221, 267-278, 284
Fraser, 69
Fraser, Charles, 26, 56, 234
Freeling, 171
Fremantle, 234
Frome, Capt., 151, 169

GARDINER, 135
Gibbu, Jimmy, 93
Gibson, Alfred, 217, 221, 283
Gilbert, 101-2
Giles, 214, 215, 216-222, 278, 282, 284, 296
Gipps, Governor, 89, 94
Gosse, W. C., 213-5, 218, 277, 278
Goyder, 170
Grant, Lieut. J., 5.
Grant, Harper, and Anderson, 213
Gray, 190 *et seq.*, 201, 202
Gregory, A. C., 102, 106, 173, 186, 189, 193, 205, 207, 210, 242, 253, 287
Gregory, Frank, 212, 242, 243, 253-263
Gregory, H. C., 173-4, 242, 243, 244, 247
Grey, Sir G., 176, 235, 237-242

HACK, Stephen, 172
Hack, 5
Hamilton, 135
Hann, Frank, 292-6
Hann, Wm., 130
Harding, 257 *et seq.*
Hardwicke, 172
Harris, J., 68, 69
Harris, Dr., 33
Harris (Babbage), 173
Hart, Capt., 137

INDEX OF NAMES OF PERSONS

Hawdon, Joseph, 135-6
Hawker, 152
Hawson, Capt., 142
Hedley, G., 225
Helpman, Lieut., 244
Hely, Hovenden, 107
Hentig, 106
Henty, 88, 137
Hergott, 178
Heywood, 152
Hindmarsh, Governor, 135
Hodgkinson, 203, 223
Hopkinson, 69, 74
Horrocks, 152-3
House, 293
Hovell, Capt., 46
Howitt, 196, 198
Hughes, Walter, 209
Hughes, 152
Hulkes, 169
Hume, H., 21, 39, 42-49, 62, 63, 67, 82, 153
Hume, K., 42
Hunter, Capt., 2

IRBY, 244

JACKY-JACKY, 114-5, 121
Jardine, Alec, 124
Jardine, Frank, 124
Jardine, John, 124
Johns, Adam, 226
Johnson, 4
Johnston, Capt., 2
Jones, G. L., 286 *et seq.*

KEKWICK, 180
Kelly, 106
Kennedy, E. B., 89, 90, 91, 110 *et seq.*, 186, 253
King (Burke and Wills), 190 *et seq.*, 202
King, Governor, 5
King, Lieut. P. P., 38, 51, 234
King, Private, 244
Kyte, Ambrose, 187

LANDELLS, G. J., 188
Landsborough, W., 200, 206-8
Lang, 4
Langbourne, 135
Larmer, 80
Lawson, Lieut. W., 10-15, 52
Leech, 283
Leichhardt, 77, 90, 95-109, 125, 207, 228, 250, 252
Leslie, P., 95
Lewis, 209 *et seq.*, 222-3
Light, Colonel, 135
Lindesay, Sir P., 78
Lindsay, David, 227, 282-284
Lockeyer, 233
Logan, Capt., 56, 58
Luff, 115
Lukin, Gresley, 224
Lushington, Lieut., 237
Lynd, R., 96, 100, 125, 131

MACLEARY, G., 68-9, 71
Macmanee, 69
Macphee, 227
Macpherson, R., 229
Macquarie, Governor, 19, 21, 30, 34, 43
Maitland, 293
Mann, J. F., 104, 109
Marsh, James, 20
Massie, 290
Matthews, 93
McKinlay, 190, 200, 201-6

McMillan, Angas, 93
Meehan, 38, 43
Meekleham, 244
Miller, 172
Mills, W. W., 290
Mitchell, Commissioner, 89
Mitchell, Sir Thos., 35, 76-92, 101, 104, 110, 136, 153, 156, 252
Mitchell (Kennedy's expedition) 120
Moore, 237
Mueller, Baron von., 216 *et seq.*, 247 *et seq.*
Mulcay, 282, 267
Mulholland, 69, 74
Murray, Sir G., 75
Myalls, 80

NEILSON and Williams, 208
Niblett, 120
Nicholson, Trooper, 287
Nicholson, Wm., 96

OAKDEN, 169
O'Donnell, 228, 281
Ord, 287, 292
Ovens, Major, 43
Overlanders, 135
Oxley, 17, 21, 22, 23-41, 43, 51, 53, 60, 68, 76, 85, 86

PALMER, 4
Pamphlet, 39
Parry, 174
Parsons, 40
Patterson, 4
Patton, 192, 198
Peron, 5
Phillip, Governor, 1-4
Piesse, 167
Poole, 156-7, 161, 170, 265
Preston, Lieut., 234
Prout, 223-4
Purcell, 198

ROBINSON, 152
Robinson (Giles) 216
Roe, 233-7, 254, 256
Roper, 102
Rossitur, Capt., 142, 149
Rudall, 288
Russell, Stuart, 95, 98

SAUNDERS, P., 226
Scarr, F., 224, 227
Scott, 52
Scrutton, 125
Scully, Capt., 243
Smith, Wm., 142
Smith (Grey) 241
Somer, 123, 205
Stanley, Capt., 112
Stanley, Lord, 89
Stansmore, 291
Stapylton, 88
Stephenson, W., 89
Stirling, 39, 40, 234, 239
Stock, 172
Stockdale, H., 228, 282
Stone, 198
Stokes, Capt., 91, 102, 238
Strzelecki, Count, 94
Stuart, 106, 156, 161, 174, 175-185, 186, 227
Sturt, Capt. 23, 35, 59, 75, 77, 80, 82, 85, 88, 103, 136, 153-168, 175, 204, 233, 253, 258
Swinden, 172

INDEX OF NAMES OF PERSONS

TATE, 130
Taylor (geologist), 130
Taylor, 120
Tench, Capt., 3, 4
Thompson, 172
Thring, 184
Throsby, 43
Tietkins, W. H., 217 *et seq.*, 222 *et seq.*, 291
Tommy (Giles) 220
Trigg, S., 253

UNIACKE, 39

VALLACK, 121
Vancouver, 233

WALCOTT, 244
Walker, Dr., 241
Walker, Frederick, 122, 200, 208
Wall, 120
Wannon, R., 137
Warburton, Major, 29, 172, 174, 178, 208-13, 214, 218, 221, 286

Warburton, Richard, 209
Warner, 131
Warrigals, 93
Welch, 196 *et seq*.
Wentworth, W.C., 10-15, 22
White, Surgeon, 2
Wickham, Capt., 58
Wild, Joseph, 43
Wells, L. A., 282, 284, 285-292
Wells, C. F., 286 *et seq.*
Wills, 122, 185, 186-200, 202
Wilson, Dr. J. B., 234
Wilson, J. S., 247
Windich, Tommy, 274
Wood, Chas., 241
Worgan, Surgeon, 4
Wright, 193
Wylie, 147-8

YOUNG, 218

ZOUCH, Lieut., 80.

INDEX OF PLACE NAMES.

ABUNDANCE, Mt., 90
Adder Water-holes, 226
Adelaide, 135, 273
Adelaide R., 184
Admiralty Gulf, 295
Albany, 242
Albany Pass, 118, 124
Albany, Port, 113
Alberga R., 215, 217
Albert R., 102, 123, 190, 200, 204, 206, 250, 290
Albury, 47
Alexander Springs, 277, 281, 290, 291
Alexandria L., 73, 138
Alfred and Marie Range, 221
Alice Springs, 209, 213, 222, 227
Alps, Australian, 47
Amadeus, L , 214, 216, 222, 291
Anson Bay, 206
Anthony Lagoon, 226
Arbuthnot Range, 36, 54
Archer R., 128
Arden, Mt. 139, 142, 154, 178
Arnhem's Land, 228
Arthur R., 227
Ashburton Range, 182, 185
Ashburton R., 221, 257
Attack Creek, 181
Augusta, L., 276, 286
Augusta, Port. 173
Augustus, Mt., 255, 257
Australia Felix, 84-88, 90
Australian Alps, 47
Australian Bight, 145-9, 270
Australian Sea (inland), 35, 40
Avoca R., 87
Ayer's Rock, 214

BALLONE R., 89, 90, 105
Barcoo R., 88-92, 104, 106, 111, 122, 173, 186 251 *et seq.*
Barlee, L., 268
Barrier Range, 158
Batavia R., 128
Bathurst, 19, 21, 92
Bathurst's Falls, 37
Bathurst, L., 43
Beagle Bay, 279
Becket's Cataract, 37
Beltana, 218
Belyando R., 253
Benson, Mt., 138
Bernier Island, 239
Berimma, 42
Birdum, 183
Blackheath, 13
Blackwood R., 243
Blanche, L., 170, 201, 203, 253
Blaxland, Mt., 14
Blue Mud Bay, 228
Blue Mts., 3, 4 *et seq.*
Bogan R., 63, 80
Bolgart Springs, 243
Bonney, L., 136, 181
Bonythorn Range, 222
Boundary Dam, 219
Bourke, 82, 89
Bowen, Port, 39

Bowen River, 205
Boyne R., 39
Braeside, 288
Brinkley Bluff, 179
Brisbane R., 39, 41
Broadsound, 206
Brodie's Camp, 181
Brown, L., 235
Brown, Mt., 208
Broken Bay, 1, 2
Bruce, Mt., 255, 257
Buchan R., 93
Buchanan's Creek, 224-5
Buchanan Creek, 228, 282
Bulloo, 192
Burdekin R., 101, 123, 124, 205, 251
Buree, 89
Burt's Creek, 210

CAERMARTHEN Hills, 1
Caledonia Australis, 94
Cambridge Gulf, 51, 278, 280 *et seq.*
Campbell R., 18, 19
Canning Downs, 55
Carnarvon Range, 274
Careening Bay, 51
Carpentaria Downs, 125
Carpentaria, Gulf, 89, 91, 101, 111, 123, 189, 251
Cassini Is., 51
Castlereagh R., 35, 64, 66
Cecil Plains, 105
Central Mt. Stuart (Sturt), 180, 209
Chambers's Creek, 174, 176, 178, 179, 182
Chambers Pillar, 179, 216
Chambers River, 183
Charlotte Waters, 215
Charnley R., 294
Chauvel's Station, 104
Claude R., 91, 106
Cloncurry R., 190, 223
Cockburn Sound, 234
Coen R., 128
Cogoon R., 90, 106
Collier B., 294
Comet Creek, 101, 206
Condamine R., 55, 95, 101, 103
Coolgardie, 289
Cooper's Creek, 89, 92, 107, 111, 165, 175, 186, 189, 196, 201, 250 *et seq.*
Corella Lagoon, 225
Cowcowing, 264 *et seq.*
Cox, R., 10
Cresswell Creek, 225
Culgoa, R., 89, 90, 112
Cunningham's Gap, 58
Curtis, Port, 39

DALY, R., 206
Daly Waters Creek, 183, 227, 280
Dampier's Land, 237
Darling Downs, 55, 88, 95, 100, 103
Darling R., 64-7, 70, 77, 78, 80, 82, 84, 87, 112, 154, 192, 253
Darlot, L., 289, 291
Davenport Range, 174. 178
Dawson R., 101, 130, 251

300

INDEX OF PLACE NAMES 301

Deception, Mt., 143
De Grey R., 258, 262, 279
Denison, Port, 205
Denmark R., 234
Depot Glen, 158 *et seq.*
Derby, 292
Diamantina R., 204, 223, 225
Dorre Is., 239
Doubtful Bay, 294
Douglas Creek, 174
Doyle's Well, 289
Dumaresque R., 54

EAST Alligator R., 206
Einnesleigh R., 125
Elder Creek, 293
Elizabeth, L., 170, 172, 174
Elsey Creek, 107-8, 250
Empress Spring, 290
Emu Is., 17
Endeavour R., 131
Escape R., 115, 129
Escape Cliffs, 206
Esperance B., 270, 279
Essington, Port, 101, 103, 250
Eucla, 270, 272
Euroomba, 251
Eva Springs, 210
Everard R., 223
Exmouth, Mt., 36
Eyre, Lake, 174, 178, 223
Eyre's Creek, 164, 190, 123

FARMER, Mt., 265
Fincke Creek, 179, 181, 210, 229
Fincke, Mt., 176
Fish R., 18
Fitzgerald R., 237
Fitzmaurice R., 247
Fitzroy R., 101, 226, 251, 278, 279, 287, 293
Fletcher's Creek, 124
Flinders Range, 143
Flinders R., 122, 190, 208
Flood's Creek, 158, 166
Flying Fox Creek, 183
Fortescue R., 256, 258
Fossilbrook, 131
Fowler's Bay, 144, 172, 176
Frances, L., 208
Fraser's Range, 283
Fremantle, 239, 256
Freeling, Mt., 180
Frew's Pond, 183
Frew R , 179
Frome, L., 151

GAIRDNER L., 172, 173, 176
Gantheaume B., 240
Gascoyne R., 221, 240, 244, 254 *et seq.* 264
Gawler Range, 172
Geelong, 48
Geographe Bay, 234
George the Fourth, Port, 238
George, L., 43, 46
Georgina R., 207, 223, 224-5
Geraldine, 254-5, 266
Gibson's Desert, 218
Gibson's Station, 178
Gilbert, R., 125, 251
Gippsland, 93
Glenelg, R., 88, 137, 176, 238, 294
Gnamoi, R., 78
Godfrey's Tank, 290
Goulburn Plains, 44, 69
Goulburn R., 47, 136
Grampian Mts., 137
Great Australian Desert, 249

Gregory, L. (Eyre), 174
Gregory R., 207
Grey Fort, 161, 164
Grose, R., 8, 10
Gundagai, 68
Gwydir R., 79

HALE R., 227
Hall's Creek, 290
Hamilton Springs, 181
Hampton Range, 272
Hammersley Range, 255, 257, 258
Hann R., 293
Hanover Bay, 237
Hanson Bluff, 179
Hardey R., 257
Harris, Mt., 35, 62, 66
Hastings R., 37
Hawdon, L., 137
Hawkesbury R., 1, 4
Hawkesbury Vale, 52
Hay R., 229, 235
Haystack, Mt., 93
Helena Spring, 290
Hopeless, Mt., 143, 151, 188, 193, 253
Herbert R., 207, 223
Hergott Springs, 178
Hermit Range, 174
Hovell R., 47
Hugh R., 181, 210
Hume, R., 44, 70, 87
Hunter R., 21, 53, 100, 103

ILLAWARRA L., 6
Impey R., 254
Inland Sea, 35, 40
Irwin R., 244
Isaacs R., 101, 104, 206
Israelite B., 270

JARVIS Bay, 38
Jervois Ranges, 227
Jimbour, 100
Joanna Springs, 29, 286 *et seq.*

KALGAN R., 235
Karaula R. (Darling), 79, 82
Katherine Creek, 183, 228
Katherine Station, 228
Kenneth, Mt., 264
Kilgour R., 227
Kimberley, 228, 278, 280 *et seq.*
Kindur R., 78
King Edward R., 294
King George's Sound, 51, 52, 144, 149, 233
King Leopold Range, 279, 292
Kintore Range, 222
Kojunup R., 243

LACEPEDE B., 138
Lachlan R., 6, 19-21, 26, 30, 35, 40, 51, 68, 69, 85, 233
Lagoons, Valley of, 101, 123
Laidley's Ponds, 156, 157
Lansdowne Hills, 1
La Trobe R., 94
Laverton, 293
Leichhardt R., 102, 202, 204, 251
Leisler, Mt., 222
Leschenhault R., 239
Limestone, 58
Lincoln, Port, 139, 142, 144, 149
Lindsay, Mt., 234
Lindsay R., 137
Little, Mt., 208

INDEX OF PLACE NAMES

Liverpool Plains, 36, 52, 78
Liverpool Range, 37
Loddon, R., 87
Lofty, Mt., 136
Logan Vale, 56
Lyons, R., 255, 257

MACALLISTER R., 94
Macarthur R., 227
Macdonnel Range, 179, 210, 214, 227, 229
Macdonald, L., 212, 291
Macedon, Mt., 88
Mackenzie, R., 101, 104, 251
Macquarie, Port, 37, 39
Macquarie R., 15, 18, 22, 30-34, 40, 63, 233
Maneroo, 43
Manning R., 38
Maranoa R., 90, 91, 110
Margaret R., 280
Margaret, Mt., 105, 269
Marshall R., 227, 229
Marryat R., 215
Mary L., 208
Massacre L., 202
McConnel, Mt., 205
McIntyre's Brook, 55
McKinlay's Range, 204
McPherson's Station, 106
Menindie, 189
Midway Well, 286
Mitchell R., 94, 127, 131, 154, 251
Monaro, 53
Monger, Mt., 283
Moran, R., 294
Moreton Bay, 39, 53, 56, 100, 247
Moore L., 244, 267
Moore R., 244
Moorundi, 156
Muckadilla Creek, 106
Mueller, Fort, 278, 283
Mueller Creek, 190, 204
Muirhead, Mt., 138
Mulligan R., 223
Murchison R., 221, 237, 245, 253 *et seq.*, 265, 267, 273, 283
Murray R., 44, 70, 83, 84, 86, 135-6, 150, 153
Murrumbidgee R., 30, 43, 46, 53, 68-70, 74, 85, 86, 153
Musgrove Range, 214

NAMOI R., 78
Napier Broome Bay, 295
Narran R., 89
Nattai, 6
Naturaliste, C., 235
Neale Creek, 178
Nepean R., 4, 6, 15, 17, 19
Newcastle Waters, 182-3, 185, 227
New Year's Creek, 63
New Zealand, 46, 53, 58, 75
Nicholson R., 94, 207
Nicholson Plains, 280
Nickol B., 256, 262
Nive R., 252
Nogoa R., 91
Norfolk Is., 58, 75
Norman R., 123, 204
Normanby R., 131
Northumberland C., 27, 88
Nundawar Range, 78

OAKOVER R., 209, 211, 213, 258, 261, 288
Oaldabinna, 218
Olga, Mt., 216
Oodnadatta, 229, 293
Ord R., 278, 280
Ovens R., 47
Oxley's Tableland, 63

PALLINUP R., 236
Palmer R., 131, 132
Pandora's Pass, 52
Peak Downs, 101
Peak Station, 221, 278
Pearce Point, 247
Peel's Plains, 55
Peel Range, 28, 29, 30
Peel R., 36
Pernatty, 172
Perth, 209, 213, 235, 241, 244, 255, 270, 273
Phillip Is., 58
Phillips Creek, 180
Phillips R., 293
Planet Creek, 101
Plenty R., 227, 229
Poole, Mt., 161
Portland Bay, 88
Powell's Creek, 226
Prince Frederick Harbour, 294
Prince Regent's River, 237, 294
Princess Charlotte B., 131
Pudding Pan Hill, 129
Pumice Stone R., 39

QUEEN Charlotte Vale, 19

RAFFLES Bay, 234
Ragged, Mt., 279
Ranken R., 224
Rannes, 251
Rawlinson Ranges, 281, 291
Red Hill, 161
Remarkable, Mt., 173
Richmond Hill, 2
Riley, Mt., 236
Rockhampton, 123, 124, 200
Rockingham Bay, 112, 123
Roe R., 294
Roebourne, 213, 226, 281
Roebuck Bay, 278
Roper R., 102, 108, 183, 250
Rossitur Vale, 142
Rudall R., 289
Russell Range, 236

SAMSON, Mt., 257
Saxby R., 123
Seaview, Mt., 37
Segenhoe, 53, 56
Separation Well, 286, 288
Serle, Mt., 151, 157
Shark's Bay, 239, 255, 265
Shaw R., 258
Shelburne Bay, 115, 119, 131
Sherlock R., 262
Shoalhaven R., 21, 43, 44, 53
Sir John Range, 293
Somerset, 124, 127, 129
South Australia, 135
Spencer's Gulf, 139-40
Squires, Mt., 283
Stansmore Range, 291
Staaten R., 126
Stephens, Port, 38
Stevenson Creek, 179
St. George's Range, 278
St. George's Rocks, 90, 91, 100
St. Vincent's Gulf, 135
Stony Desert, 162
Strangways Creek, 183
Strathalbyn, 123
Streaky Bay, 139, 144, 172
Strzelecki Creek, 162, 165, 253
Sturt's Creek, 210, 249 *et seq.*, 287, 291
Sutton R., 101
Swan Hill, 87
Swan R., 140, 153, 234, 237, 264
Swinden's Country, 172, 175

INDEX OF PLACE NAMES

Tambo R., 93
Tate R., 131
Tench R., 4
Tennant's Creek, 181, 224
Termination Hill, 40
Thistle Cove, 149
Thompson's Station, 178
Thomson R., 107, 108, 111, 122, 206, 252, 253
Timor, 237
Torrens, L., 135, 139, 140, 151-3, 156, 161, 169, 173, 174, 178, 201, 253
Tumut R., 68
Tweed R., 39

Vanstittart Bay, 295
Victoria (Port Essington), 103
Victoria, 84
Victoria, Lake, 94, 156
Victoria R. (Barcoo), 91, 110-2, 180, 205, 247 et seq., 280
Victoria Spring, 220, 283

Walsh R., 131
Warburton Creek, 213
Warburton Desert, 289
Warburton Range, 293
Warning, Mt., 56
Warragamba R., 8, 10, 14
Warrego R., 91, 107, 112, 208
Waterloo Wells, 210

Weathered Hill, 170
Welbundinum, 284
Welcome. Mt., 266
Weld Springs, 274, 276
Wellbeing, 245
Wellington Valley, 33, 62
Western Port, 46, 94
Weymouth Bay, 114, 119, 121
Whaby's Station, 68
Williams R., 239
Williora R,, 154, 157
Williorara, 167
Wimmera R., 88, 137
Windich Springs, 274
Wingillpin, 175
Woodhouse Lagoon, 290
Woolloomooloo, 50
Wyndham, 294

Yarraging, 268
Yass Plains, 43, 84
Yilgarn. 235
York, C., 112, 123
York, Mt., 14
Yorke Peninsula, 222
Youldeh, 218
Yuin, 273
Yule R., 262

Zamia Creek, 101